LOW INTENSITY OPERATIONS

LOW INTENSITY
OPERATIONS

Subversion, Insurgency, Peace-keeping

FRANK KITSON

Faber and Faber
3 Queen Square
London

First Edition 1971
Published by
Faber and Faber Limited
3 Queen Square
London W.C.1

Reprinted 1972, 1973 and 1975

Printed in Great Britain by
Redwood Burn Limited
Trowbridge & Esher

ISBN 0 571 09801 0

Contents

v

Organizational charts

Acknowledgements

Many people have helped me in the preparation of this book and although I cannot mention them all by name I am none the less most grateful to them. In a few cases my debt is such that I feel bound to record it here, and in this respect I should like to start by expressing my most sincere thanks to the Master and Fellows of University College Oxford for allowing me to live amongst them as a member of their Senior Common Room for the year in which I have been writing the book. This has been a great privilege to me, and I only hope that it has not been too great an imposition for them.

Next I wish to thank Professor N. H. Gibbs, Chichele Professor of the History of War at Oxford University for supervising my study. His tact as an adviser together with his academic knowledge and integrity are responsible for the results of my work being publishable, and I am very conscious of my good fortune in having had him as a guide. Finally I should like to mention the many officers of the British and United States Armies who helped me during my visits to their various schools, colleges, units and establishments. Without exception they gave freely of their time and experience and I hope that I have done justice to their views.

University College   1st September 1970
Oxford

Foreword

General Sir Michael Carver, GCB, CBE, DSO, MC, ADC
Chief of the General Staff

Nobody could be better qualified than Brigadier Frank Kitson to write on this subject. He has had a wide experience both of operations and intelligence against terrorists and in the different field of peace-keeping. In Malaya, Kenya and Cyprus he approached the problems of this unfamiliar type of warfare, if it can be called that, with a combination of determination, unprejudiced objectiveness, devotion to the task and high personal courage. I myself had first hand knowledge of his exceptional skill in this field both in Kenya and in Cyprus.

His approach could not be better defined than in his own words at the end of Chapter 6, where he says:

'The process is a sort of game based on intense mental activity allied to a determination to find things out and an ability to regard everything on its merits without regard to customs, doctrine or drill.'

The necessity for the intimate integration of intelligence and operations is his most important lesson and the one least appreciated by the conventional soldier. Frank Kitson's great virtue is that he is above all a realist, in spite of being both an idealist and an enthusiast. The reader will not find in these pages a purely academic theoretical exercise. He will, however, find some stimulating and original suggestions about the tasks which confront the Army in the field of 'low intensity operations' and about the methods which should be used both to prepare for and execute them.

This book is written for the soldier of today to help him prepare for the operations of tomorrow. It will be of the greatest possible help to him, and I hope it will be read by all those concerned with training the Army.

Introduction

During the twenty-five years which elapsed between 1945–70 the British Army took part in a large number of operations of one sort or another. Julian Paget mentions thirty-four in a book[1] which he wrote on this subject, and there are now one or two more to add to his list which only goes up to 1966, Anguilla and Northern Ireland being cases in point. Of all the operations which he quotes, only four could be described as Limited War, i.e. Korea, Suez, the move into Kuwait in 1961, and the Indonesian confrontation. All the rest were concerned with countering subversion or insurgency, or with peace-keeping operations.

The size and organization of the army during this period was not of course tied exclusively to the operations in which it became involved. On the contrary, one of the main factors governing its composition was the requirement to maintain a force in Europe which would appear to the world at large as a convincing contribution to a credible deterrent. But the army's potential commitments were, for most of the period, very much greater than they are today, its size was correspondingly larger, and as a result there was enough spare capacity at any given moment to handle such emergencies as did in fact crop up. Furthermore, the margin was wide enough for a certain amount of clumsiness to be compensated for by sheer weight of numbers, although even in this period of comparative strength Britain sometimes fell short of her agreed contribution to NATO as a result of the demands made by operational requirements. During the 1960's Britain made far-reaching cuts in her commitments and at the same time reduced the size of her forces, and this process is still going on. But as the overall size of the army is reduced so the built-in margin for dealing with the unexpected sinks. Although this can be offset to some degree by including extra units in the order of battle, the extent of such an allowance in these days of financial stringency is unlikely to equal the cushion which automatically existed when a large conscript army was in being as it was in the 1950's, backed by the forces of the colonies and dependencies. As a result it will be even more important in the future for the army to handle its tasks in the

[1] JULIAN PAGET, *Counter-Insurgency Campaigning*, Faber and Faber, 1967, p.180.

1

most efficient way possible. It will be useless to rely on sheer weight of numbers to put right mistakes made because of a lack of proper preparation.

The purpose of this book is to draw attention to the steps which should be taken now in order to make the army ready to deal with subversion, insurrection, and peace-keeping operations during the second half of the 1970's. The book is slanted towards the situation and needs of the British army in so far as its outward form is concerned, but the analysis of past campaigns and the prediction of the likely nature of future operations from which the specific recommendations are made, is relevant to the armies of most countries, as indeed are many of the recommendations themselves, including all of the important ones.

In writing on this subject one of the most difficult problems concerns the matter of terminology. The British Army gives separate definitions of Civil Disturbance, Insurgency, Guerilla Warfare, Subversion, Terrorism, Civil Disobedience, Communist Revolutionary Warfare, and Insurrection on the one hand and of Counter Insurgency, Internal Security, and Counter Revolutionary Operations on the other. Elsewhere conflicts are variously described as Partisan, Irregular or Unconventional Wars, and the people taking part in them have an even wider selection of labels attached to them. Furthermore, although a particular author will use one of these terms to cover one aspect of the business and another to cover another, a different author will use the same two terms in a totally different way. Under the circumstances any attempt to re-define all the terms is more likely to bring confusion than enlightenment, but it is none the less important to make an attempt to explain the more important ones used in this book. No doubt the most satisfactory answer would be to settle for one all-embracing expression which would cover every form of conflict carried out by people other than those embodied in the legal armed forces of a country. The nearest approach to such a term would perhaps be 'Revolutionary Warfare' but to many people such an expression is too heavily weighted towards the activities of communist or left-wing groups: it would somehow seem wrong to describe the activities of Grivas or Mihailovic in this way. Another possible solution would be to take a well-known term such as subversion or insurgency and to define it arbitrarily in such a way as to cover the whole subject. This would be perfectly reasonable so far as the meaning given to the words in the dictionary is concerned, but in each case they have become accepted as applying to one particular part of the business: to make one of them apply to the whole would lead to confusion.

For this reason the best course would seem to be to define subversion and insurgency in a way which accords with modern practice, and to leave the reader to interpret other terms in the light of the text. Subversion, then, will be held to mean all illegal measures short of the use of armed force taken by one section of the people of a country to overthrow those governing the country at the time, or to force them to do things which they do not want to do. It can involve the use of political and economic pressure, strikes, protest marches, and propaganda, and can also include the use of small-scale violence for the purpose of coercing recalcitrant members of the population into giving support. Insurgency will be held to cover the use of armed force by a section of the people against the government for the purposes mentioned above. This would seem to be the sense in which the terms are understood by British authorities such as Sir Robert Thompson.[1] Naturally, subversion and insurgency can take place in the same country at the same time, and either or both can be supported by a foreign country, which may well provide the impetus. Between them these terms cover virtually every form of disturbance up to the threshold of conventional war. There is, however, one loophole concerning disorders which are not aimed at overthrowing the government or even at forcing it to do something which it does not want to do, and this relates to activities which might take place as a protest against the actions of some non-government body, or possibly as a demonstration of solidarity with a group or cause elsewhere in the world. The handling of such situations would in British Army parlance come under the heading of Internal Security Operations, a term which also covers the suppression of various forms of subversion and insurrection as well. In practice this form of incident is not sufficiently significant from a military point of view to warrant a definition of its own, and no further distinction between it and subversion or insurgency will be made in this book.

One other term which merits a definition is peace-keeping which

---

[1] ROBERT THOMPSON, *Defeating Communist Insurgency*, Chatto and Windus, 1967, p.28.

3

will be used in this study to mean preventing by non-warlike methods, one group of people from fighting another group of people. Peace-keeping does not involve the activities of an army which formally attacks one or both parties to a dispute in order to halt it, because although this might be done with a view to re-establishing peace, the activity itself would be a warlike one and would be of a totally different nature to a peace-keeping operation.

Perhaps at this stage it is worth trying to identify the way in which subversion and insurgency differ from other forms of war. One of the main differences arises directly out of the definitions given, which makes it clear that both are forms of civil conflict because both involve action by one section of the population of a country against another section. This is true even when the main impetus comes from outside. It is also true when the governing authority is an occupying power because it is virtually certain to have the support or acquiescence of some of the indigenous population at the least. A more important difference lies in the relationship which exists between the use of force and the use of other forms of pressure. The people of a country can only be made to rise up against the authorities by being persuaded of the need to do so, or by being forced into doing it. Usually those involved in organizing subversion envisage replacing the authorities ultimately, and ruling in their stead, and when this point is reached it is better to have people who are giving their support willingly. In other words, in theory at least, insurgents are likely to use persuasion on the people whose support they want, and only use violence to back it up if necessary: in practice insurgents sometimes use force at the wrong time because of errors of judgement, bad temper or an inability to control their followers. If the organizers of the campaign can obtain the support of a large enough proportion of the population, and demonstrate the fact to the government by such means as strikes and protest marches, they may be able to persuade the government to give in without using force at all. But if the government has an appreciable hold on the population, or if it derives its authority from an occupying power which is determined to stand fast, then force will be needed for attacking the government's forces, for defending those involved in the subversion, and for creating economic difficulties. Therefore the second main characteristic of subversion and insurgency is that force, if used at all, is used to reinforce other forms of persuasion, whereas in more orthodox forms of war, persuasion in various forms is used to back up force.

One well-known author describes subversion and insurgency, which he lumps together under the name of modern warfare, as being an interlocking system of actions, political, economic, psychological and military that aims at the overthrow of established authority in a country.[1] Mao Tse Tung, who is probably the most highly regarded authority on the subject, envisages the government being overthrown by a regular revolutionary army, which develops from, and operates with guerilla troops after a protracted campaign during which the population of the country concerned is methodically indoctrinated and organized to support the cause. General Grivas, on the other hand, adopted a totally different approach in the Cyprus campaign. He understood from the start that he could not achieve his aim by developing guerillas into regular troops. In his preliminary plan for the Insurrectionary Action in Cyprus which he drew up in Greece before his departure he stated:

> 'It should not be supposed that by these means we should expect to impose a total defeat on the British forces in Cyprus. Our purpose is to win a moral victory through a process of attrition, by harassing, confusing and finally exasperating the enemy forces with the effect of achieving our main aim . . .'[2]

Earlier in the plan Grivas had laid down that the campaign was to consist of action designed to draw the attention of international opinion to the Cyprus question so as to mobilize international diplomacy.[3] The mixture of harassing the government and mobilizing international opinion is a theme that constantly recurs. In Algeria an attempt was made to develop a full-scale regular force to attack the French army in the field but it was rapidly abandoned because of the heavy casualties incurred.[4] The final political victory gained by the National Liberation Front was the result of what Grivas referred to as harassment and it

---

[1] ROGER TRINQUIER, *Modern Warfare*, Pall Mall Press, 1964, p.6.
[2] GEORGE GRIVAS, *Guerilla Warfare*, Longmans, 1964, p.92.
[3] Ibid., p.91.
[4] EDGAR O'BALLANCE, *The Algerian Insurrection*, Faber and Faber, 1967, pp.143 and 154.

came about at a time when the insurgents were all but beaten militarily.[1]

Although there is a wide difference of approach between one exponent of insurgency and another, there is none the less a dividing line of a sort between it, and conventional war. But even this is really no more than a matter of emphasis. If subversion fails to achieve the aim, it merges imperceptibly into insurrection, which at one end of the scale covers the activities of small sabotage or terrorist groups but which spreads across the operational spectrum to include the activities of large groups of armed men. If these gangs become sufficiently numerous and well-armed to take on the forces of the government in open combat on relatively even terms, insurgency merges into orthodox civil war, because at this stage force has again become the senior partner. Such a situation could arise if, for example, a significant number of government troops were to defect to the enemy, or if the insurgents were supported from outside the country. Should the outside support include the operation of foreign troops, the civil war would merge into a normal international war which could easily be waged in conjunction with civil war, insurgency and subversion. It is not easy to cover every set of circumstances by exactly defined terms, nor in the last resort is it necessary to do so. The purpose of this book is to consider the action which should be taken in order to make the army ready to deal with subversion, insurrection and peace-keeping, but the army has got to be ready to deal with its other tasks as well. If there is an overlap between the busier end of insurgency and normal war it is of little consequence, providing that men are there who know how to cope with the situation.

Having tried to show in very broad outline how subversion and insurgency differ from what are generally considered to be more conventional forms of warfare, it is now necessary to hark back to the stated purpose of the book in order to stress that it is the army's contribution which is being studied, with a view to identifying the steps which should be taken in advance to prepare it for its task. Although this falls short of being a comprehensive coverage of the whole field, it is none the less an ambitious project. It could be argued that any worthwhile coverage of it should

[1] EDGAR O'BALLANCE, *The Algerian Insurrection*, Faber and Faber, 1967, p.220.

include a survey of the theory of this sort of war, together with an analysis of past campaigns and a study of the way in which all parts of the government machinery combine together in pursuit of the objective, and these subjects will certainly have to be discussed to some extent. But if they were to be covered in any sort of detail the book would be so long that only a dedicated student of insurgency would read it, and he would probably be an insurgent. It is unlikely that it would be read by many of those who are now in a position to prepare the army for future operations. Furthermore it would be difficult to concentrate attention on precautionary measures if they were to be depicted against such an extensive background. These matters are only discussed, therefore, to the extent necessary for the book to achieve its stated purpose, and the same consideration has been employed in deciding on the amount of detail to be included about the military operations themselves. In every case the deciding factor has been the extent to which understanding is relevant to preparation.

As the book progresses it will become apparent that the army's contribution to fighting subversion and insurgency usually falls under one of two headings. In the first place the army has got to provide units which are trained, organized and equipped to carry out the sort of operations given to them, and in the second it is responsible for producing properly educated commanders and staff officers capable of advising the government and its various agencies at every level on how best to conduct the campaign. In this connection it is worth pointing out that as the enemy is likely to be employing a combination of political, economic, psychological and military measures, so the government will have to do likewise to defeat him, and although an army officer may regard the non-military action required as being the business of the civilian authorities, they will regard it as being his business, because it is being used for operational reasons. At every level the civil authorities will rightly expect the soldier to know how to use non-military forms of action as part of the operational plan, although once it has been decided to use a particular measure they will know how to put it into effect. This point is not always understood by soldiers whose recollections of fighting insurgency usually start at the point where they arrived in a district to find that the local administrator and policeman knew all about the business whereas they knew nothing. But this merely represents

7

an accident of timing, and reflects the fact that the administrator and policeman, being permanently stationed in the country, had learnt from experience over two or three years what the soldier should have been taught before he arrived. At the start of the trouble no one would have known what to do, and whereas there is no reason why the policeman or administrator should have known, there is no excuse for the soldier having been ignorant.

For ease of reference this book is written in three parts. Part I starts by analysing the way in which force is likely to be used in the future with a view to showing why it is necessary for the army to be prepared to fight subversion and insurgency and to take part in peace-keeping operations. Part I also explains the background to these activities in sufficient detail for subsequent chapters to be intelligible. In Part II the army's contribution is discussed under the two headings already mentioned, related to different phases of operations. For example, one chapter deals with the army's contribution during the period before trouble actually breaks out, another deals with the phase of non-violent subversion and a third with insurgency. Part III collects together requirements arising from the other two parts of the book and presents them as a series of recommendations for future action.

One final matter which requires mentioning in this introduction, concerns the moral issues involved in preparing to suppress subversion. Many regard subversion as being principally a form of redress used by the down-trodden peoples of the world against their oppressors, and feel, therefore, that there is something immoral about preparing to suppress it. Undoubtedly subversion is sometimes used in this way, and on these occasions those supporting the government find themselves fighting for a bad cause. On the other hand subversion can also be used by evil men to advance their own interests in which case those fighting it have right on their side. More often, as in other forms of conflict, there is some right and some wrong on both sides, and there are high minded and base people amongst the supporters of both parties. Fighting subversion may therefore be right on some occasions, in the same way that fostering it might be right on others, and the army of any country should be capable of carrying out either of these functions if necessary, in the same way as it should be capable of operating in other forms of war. In a democratic country it is the duty of soldiers to know how to wage war in any of its forms, and it is the duty of the people to elect representatives who will only make war when it is right to do so. When conflicts occur, soldiers like other people, have to have faith in the moral rectitude of their government to some extent, because it is not usually possible to know enough of the facts to make an absolute judgement as to the rights and wrongs of the case. But if any man, soldier or civilian, is convinced that his country is wrong he should cease to support it and take the consequences. The fact that subversion may be used to fight oppression, or even that it may be the only means open for doing so, does not alter the fact that soldiers should know how to suppress it if necessary. Moral issues can only be related to the circumstances of a particular case, and then they must be faced by soldiers and civilians alike on moral grounds.

PART ONE

TRENDS AND BACKGROUND

Chapter 1

Future Trends in the Use of Force

The purpose of this chapter is to show why it is necessary for the army to be ready to suppress subversion and insurgency, and to take part in peace-keeping operations during the second half of the 1970's. On the face of it such justification might appear to be unnecessary on the grounds that the need is obvious in relation to world conditions, with particular reference to the events of the last twenty years or so: at the time of writing both the British and United States armies are heavily engaged in these activities in Northern Ireland and Viet Nam respectively. But it can be argued that the recent past has been exceptional, that Northern Ireland and Viet Nam will both be settled within five years, and that with the proposed withdrawal of all but a small remnant of the British Army into Europe, the requirement to fight insurgents or to take part in peace-keeping operations will cease. Since this view is held by some influential people both inside and outside the army, it is necessary to make out a case for being ready to take part in these operations. The case is also relevant to those who accept the need in principle, but who are uncertain as to the relative importance which should be attached to preparations designed to fit the army to take part in these activities as opposed to preparations concerned with making it ready to fight in a conventional war.

There are two approaches towards making out a case for preparing to wage a particular form of warfare, and they apply as much to fighting conventional war as they do to countering subversion or taking part in peace-keeping operations. The first approach is to make specific predictions as to situations which are likely to arise based on foreseeable threats. The second is to show that although specific outbreaks can not be foreseen, the particular form of war in question is likely to occur in circumstances which make it probable that the country will become involved in countering it. A good case made along the lines of the first approach stands the best chance of carrying conviction, and at certain times there can be no doubt that this is the right way of tackling the problem. For example from 1937 until the outbreak of war it was clear that Germany constituted a serious and imminent menace to Britain, and it would have been foolish to have based the case for re-armament on anything other than the German threat. At other times the danger is not so overwhelmingly obvious and a choice can be made as to which approach should be used. In terms of the subject matter of this book an option of this sort exists now. It would be perfectly possible to concentrate on the threat posed by Russian inspired subversion and insurgency, and there is good material available which could be used to support a case made on these lines. On the other hand this threat is not so clear cut as that posed by Hitler in 1937 and too much emphasis on it tends to provoke the reaction that communists are being seen under every bed. If expensive preparations over a number of years had been based on making the country ready to counter this particular threat, it would probably be wise to justify continued precautions on the strength of it, in the same way that Britain's conventional war capability is justified by reference to the threat of a Russian invasion of Western Europe which has not been imminent for at least a decade. But as neither the threat of Russian inspired subversion, nor the need to prepare for it, has been regarded as being of any great importance during recent years, and bearing in mind that most of the counter-insurgency campaigns waged by the British since 1945 have not been concerned with fighting communists, it would seem better to base the case for future preparations on an analysis of world trends showing that subversion and insurgency are current forms of warfare which the army must be ready to fight, than to single out a communist threat which may not develop.

* * * * * *

There is nothing new about subversion or insurgency. Writing in the *Encyclopaedia Britannica,* Robert Asprey has this to say about Guerilla War, using the term in roughly the same way as insurgency is defined in this book.

'Guerilla warfare by tradition is a weapon of protest employed to rectify real or imagined wrongs levied on a people either by a foreign invader or by the ruling government. As such, it may be employed independently or it may be used to complement orthodox military operations. . . . In either capacity the importance of its role has varied considerably through history'.[1]

[1] *Encyclopaedia Britannica,* 1969, p.1002.

The important point to notice is that guerilla war is described as a traditional form of conflict, and that it has been used throughout history either independently or in conjunction with orthodox operations. In fact, comments on the conduct of such operations were included in a book written as long ago as the fourth century B.C. by the Chinese general Sun Tzu,[1] which still makes sense and which Mao Tse Tung is known to have studied when formulating his own ideas on the subject.[2]

But although subversion and insurgency have been known for such a long time, and allowing for the fact that their importance has varied greatly over the centuries, it seems that they have seldom been used to better effect than they have in the past twenty-five years. From one end of the world to the other campaigns of this sort have proliferated to such an extent that some commentators now talk about it as 'Modern Warfare'[3] and mention is even made of a new dimension being introduced into conflict. Whether or not the situation justifies the terminology is relatively unimportant. What obviously does matter is the extent to which this trend in the use of force will continue, and in order to assess the likelihood of its doing so, it is necessary to examine the reasons for it having developed in the way it has during recent years.

It is possible to identify three separate factors as being responsible for the rise in the incidence of subversion and insurgency. The first of these concerns the changing attitude of people towards authority. The second relates to the development of techniques by which men can influence the thoughts and actions of other men. The third factor is the limitation imposed on higher forms of conflict by the development of nuclear weapons. None of the factors on its own would account for the increase in the use made of subversion or insurgency in recent years, and by the same token some subversion or insurgency would almost certainly have taken place as it has done over the centuries, without all three of the influences being present at the same time. The existing situation could perhaps be regarded as a freak one in the context of history, because the chances of all three of the factors coming into play at

[1] SUN TZU, *The Art of War*, Translated by S. Griffiths, Oxford University Press, 1963.
[2] S. GRIFFITHS, Translator's notes to Mao Tse Tung's *Guerilla Warfare*, Cassell, 1962, p.29.
[3] ROGER TRINQUIER, op. cit., pp.6–9.

the same time must be remarkably small. But that is what has happened, and whether the situation is a freak one or not, the present generation has got to accept it. In order to decide whether these conditions will continue throughout the 1970's it is necessary to look in more detail at each of the factors in turn.

Of these three factors the question of peoples' attitudes is the most difficult to analyse and explain. Liddell Hart recognized its importance in relation to the incidence of subversion and insurgency at least eight years ago when he wrote:

'Campaigns of this kind are the more likely to continue because it is the only kind of war that fits the conditions of the modern age, while being at the same time well suited to take advantage of social discontent, racial ferment and nationalist fervours.'[1]

For the first fifteen years after the end of the Second World War, nationalistic fervour in the context of freedom from colonialism was the most usual cause of uprisings, and twelve of the twenty examples quoted as 'World-Wide Insurgencies' by Julian Paget[2] come under this heading. For the future it may well be that social discontent and racial ferment will be more important, and disturbances arising out of dissatisfaction with society, often allied with racial problems which have not yet been mastered, are already commonplace. There is no doubt that Russia has exploited these influences wherever possible and that she has done her best to foster them as a means of weakening the will of certain countries to resist the spread of communism,[3] but her direct intervention probably accounts for no more than a small proportion of the trouble. John Galtung writing in *Survival*,[4] points out that conflict is bound to increase as a result of a world breakdown in homogeneity, the breakdown of the feudal order, and peoples' reaction to the future. Although stated baldly these three reasons may sound a bit esoteric, the article itself covers a wide variety of contemporary developments and is not easy to fault. There are of course dozens of theories to account for the unsettled state of the world, some of which are infinitely more convincing than others.

[1] B. H. LIDDELL HART, Foreword to Mao Tse Tung and Che Guevara: *Guerilla Warfare*, Cassell, 1962, p.xi.

[2] JULIAN PAGET, *Counter-Insurgency Campaigning*, p.18.

[3] IAN GREIG, *Assault on the West*, Foreign Affairs Publishing House, 1969, p.2.

[4] JOHN GALTUNG, 'Conflict as a Way of Life', *Survival*, Jan 1970, p.14.

It could even be argued that the world is always in an unsettled state and that the present situation is not exceptional in any way. But whether or not the world is more unsettled than usual, it is difficult to think of good reasons why the situation should improve in the next few years, and it is most unlikely to do so in time to relieve the incidence of subversion and insurgency to any appreciable extent during the coming decade.

The second factor which concerns the techniques by which men can influence the thoughts and actions of other men is much easier to discuss. Whether or not there is more discontent in the world than was formerly the case, there is no doubt whatsoever that the means of fanning it and exploiting it are infinitely greater than they used to be, because of the increase in literacy and the introduction of wireless and television sets in large numbers. From one side of the world to the other the organizers of subversion have access to the people through these means and although the same channels of communication are available to those involved in protecting the existing order, they seldom manipulate them so skilfully as their opponents.

There are two aspects to the business of using the communications media for spreading subversion. In the first place there is the obvious one directly concerned with the progress of a particular campaign which covers the production of news sheets by illegal printing presses and the making of broadcasts by illegal wireless stations. These activities form a most important part of any subversive campaign, particularly in the early stages when the population is being mobilized to support the cause. Sometimes when the impetus for subversion comes from a foreign power or when a foreign power is in sympathy with the cause, the organizers are allowed to make use of broadcasting facilities in the friendly country concerned. For example Radio Hanoi broadcasts for the benefit of the Viet Cong, Radio Athens put out propaganda for the benefit of EOKA during the Cyprus emergency,[1] and Radio Cairo, Taiz and Sana broadcast on behalf of the insurgents in Aden and the Western Aden Protectorate.[2]

But there is another aspect to the way in which the means of mass communication are being used which concerns the general conditioning of people throughout the world to accept subversive

---

[1] JULIAN PAGET, *Counter-Insurgency Campaigning*, p.120.
[2] JULIAN PAGET, *Last Post in Aden*, Faber and Faber, 1959, p.30.

ideas so that they will act on them when the time is ripe. A very large contribution in this direction is made by Russia in her efforts to spread communism, vast quantities of books, pamphlets and magazines being used in addition to an extensive broadcasting programme. Other countries involved in similar activities include China, Cuba and Egypt. The first six chapters of Ian Greig's book[1] give an excellent survey of the situation and should certainly be read by anyone who is sceptical about the serious nature of the threat.

Turning to the future there is no doubt that from a mechanical point of view the ability of men to influence each other by the printed and broadcast word will increase, as more and more people learn to read, and as small cheap wireless sets become available in even greater quantities. There is of course no technical reason why this should work to the advantage of the organizers of subversion since their propaganda could be nullified by more effective propaganda put out by the other side. But this would involve more thought, effort, and money being devoted to the purpose than has usually been the case in the past. Whereas it would be perfectly possible to stem the tide of subversive propaganda, it would require a great deal of optimism to predict any great swing in favour of those whose business it is to protect the existing order during the next ten years.

The third factor is the limitation imposed on higher forms of conflict by the development of nuclear weapons. In its simplest form this results from the fact that the two major world powers cannot afford to risk overt warlike operations against each other, because of the ability each has for destroying the other. It has also affected the behaviour of many of the other nations of the world, because America or Russia has been obliged to use its influence to damp down any conflict which might have escalated in such a way as to bring about a threat of war between them. It would be wrong to suggest that the nuclear balance has made all orthodox wars impossible. Several have taken place in recent years, such as the India-Pakistan conflict of 1966, and the Arab-Israel war of 1967, but these appear to have been exceptional cases, the first of which afforded little threat of escalation and the second of which was quickly stopped largely as a result of the influence of the great

[1] IAN GREIG, op. cit., pp.1–70.

powers. It seems to many people that the nuclear balance has helped to limit the number of wars which have taken place since 1950 and while the balance holds it is likely to continue doing so.

Unfortunately the same limitations do not apply to subversion and insurgency, where the danger of escalation is very much less real. The communist countries well understand this. Russia, China, and Cuba, in particular, openly encourage what they like to call wars of national liberation which is to say any form of subversion or insurgency carried out in such a way as to advance their interests. In practice it is not only communists who encourage such activities and Egypt's record during the last decade is second to none. The fact is that most countries which would formerly have been prepared to go to war in pursuit of a particular interest, would now be prepared to pursue it by encouraging subversion or insurgency, providing that they knew how to do so. Furthermore the nuclear balance not only makes it necessary for countries to pursue their interests in this way in many cases, but it also makes it safer to the extent that it inhibits the country being subverted from retaliating in an orthodox way. It therefore enables a weak country to take on a stronger country to an extent which would not formerly have been possible. For both these reasons the nuclear balance has tended to increase the incidence of subversion and insurgency.

The question of how long this particular factor will continue to exert an influence depends on the time during which the balance will last. If either Russia or America were to achieve a technological breakthrough which would enable one of them to destroy the other without risking destruction in return, or if a third power of equivalent strength were to arise, a changed situation would exist. It is possible that the new circumstances like the present ones would continue to limit the opportunities for waging orthodox war and that subversion and insurgency might prosper as a result. On the other hand this might not be the case. It is not easy to know whether Russia or America will be in a position to gain a decisive lead over the other during the 1970's but recent developments indicate that both countries are prepared to continue spending money on research in order to ensure that the other does not get a lead. It is therefore fair to say that it is rather more likely that the balance will hold, than that it will be upset during the period. The fact that both America and Russia are prepared to continue spending money in this way may result in them impoverishing themselves to some extent, and it will certainly mean that less money is available in the world as a whole for relieving want and for developing backward areas, but this can only improve the chances of subversion being successful. The likelihood of the nuclear balance being seriously upset by the emergence of a third world power during the next ten years, depends on the speed at which those countries who now have nuclear weapons can improve them, and the extent to which countries now without them can develop them. In the distant future there is a possibility of China becoming a significant nuclear power, or perhaps of Europe, united in defence terms, or in some other way, achieving an equally powerful place, but there seems little chance of either of these happenings upsetting the nuclear balance during the next decade, although either America, or Russia, or both might become so obsessed by the threat posed by China that they started to neglect Europe as a result. A lessening of the influence of the great powers in Europe could make the situation there less stable and it might possibly result in the outbreak of insurgency in areas which now appear to be completely peaceful.

On balance therefore, consideration of the three major factors indicates that there is little reason to expect a reversal of the trend towards subversion and insurgency which has been such a marked feature of the last twenty-five years. Even if one or other of the factors started to operate less strongly, and the trend did go into reverse, it would not necessarily mean that any sudden change would follow. It is more likely that cases of orthodox war would become more frequent, and that cases of insurrection would gradually become less. But it could equally well be argued that a more likely situation is for the trend to continue in the present direction and reach a further stage in which the insurgents' aim is achieved before subversion becomes insurgency or before insurgency develops into full-scale civil or limited war as happened in Viet Nam during 1967–8. This could come about as a result of a further deterioration in peoples' attitude towards authority, or if those conducting the campaigns became even more adept at handling the propaganda media and combining it with other forms of subversion such as the application of economic pressure. Similarly, if the defenders of the existing order themselves become more efficient at countering subversion and insurrection, they

will be able to achieve their aim before the campaign can develop
into one of the later stages. The R.U.S.I. Journal of December
1969[1] carries an article which comes to the conclusion that low-
level urban insurgency combined with propaganda and economic
pressure, is likely to be the most popular form of operation in the
future, but it is too early to know whether this prediction will be
fulfilled.

So far, this assessment has concentrated on discussing general
trends in the world at large and it is now worth considering how
far the various countries whose interests are likely to run counter
to those of Britain understand the nature of subversion and the
uses to which it can be put in the promotion of their national
designs. From the earliest days of recorded history the stirring up
of subversion in an enemy country has been regarded by some as
an adjunct or an alternative to other sorts of operations. Sun Tzu
specifically states that conventional war should only be used if
the enemy can not be overthrown by the activities of spies and
agents sowing dissension and nurturing subversion.[2] Lawrence
of Arabia saw the fostering of insurrection as a method of
carrying out operations against an enemy army,[3] and Mao Tse
Tung in his celebrated essay 'Guerilla Warfare' stated:–

'Guerilla operations must not be considered as an independent form of warfare. They are but one step in the total war . .'[4]

There is ample written evidence to show that Chinese and Cuban
leaders understand the potentialities of this sort of war, and although the Russians are less disposed to extol the virtues of
armed insurrection, there is not the slightest doubt that they
understand the potentialities of it, and are ready and able to
foster or exploit it whenever they consider that their interests
would be served by doing so.

In fact Russia and her European satellites have already gone
some way towards subverting the countries of Western Europe,
and any military adventures which they may contemplate in the
area would almost certainly be designed to take advantage of the

---

[1] PETER DE LA BILLIERE, 'Changing Pattern of Guerilla Warfare', R.U.S.I.
Journal, Dec. 1969.
[2] SUN TZU, The Art of War, p.39.
[3] T. E. LAWRENCE, Seven Pillars of Wisdom, Jonathan Cape, 1935, pp.188–196.
[4] MAO TSE TUNG, Guerilla Warfare, translated by S. Griffiths, Cassell, 1962, p.31.

work which they have done in this respect. Operations would probably be preceded and accompanied by disruption on a sufficiently widespread scale to ensure that troops required for fighting the conventional battle would have to be diverted to deal with it. Ian Greig gives much concrete information on how subversion is being organized now. He explains for example how the fostering of it is an essential function of Soviet Intelligence,[1] how a special section of the National People's Army of East Germany, working closely with Russian Intelligence, is organized to carry out subversion in West Germany; how as long ago as 1961 East Germany had approximately 16,000 agents in West Germany,[2] and how in the early 1960's arrests of suspected communist bloc agents there were running at over 2,000 a year.[3] If these facts are looked at against assessments of the amount of disruption which a few well-trained terrorists could cause under suitable conditions, some idea of the magnitude of the problem can be obtained. A leading French writer illustrates this well in relation to two areas in France.[4]

Needless to say the organization of subversion is not restricted to Europe. Russia, the satellites, China, Egypt and Cuba amongst others have devoted a great deal of effort to subverting countries all over the world, and between them they have a considerable potential for capitalizing on their efforts should they want to do so, although the difficulties of sparking off insurrection at the right moment in relation to an overall plan, and then controlling its development should not be underestimated. Both in Europe and outside it, the question of whether a dispute manifests itself as subversion, or as insurrection, or as orthodox war, or as a mixture of two or three of these forms, is likely to depend solely on the merits of the case in relation to the particular aim being pursued at the time. For example it is perfectly possible that the Russians would have launched a campaign of subversion against the Dubcek government in 1968 had they thought that the Czechoslovakian army would have fought their invasion forces, but having decided that there was no likelihood of such an eventuality, and realizing that there was no chance of escalation to

[1] IAN GREIG, op. cit., p.95.
[2] Ibid., p.78.
[3] Ibid., p.70.
[4] ROGER TRINQUIER, op. cit., p.24.

nuclear war, they decided to use a swifter method and one which stood less chance of getting out of control. It is difficult to know in advance what form a war is going to take. All that can be said is that the general trend in the use of force is for conflicts to be fought at the subversion end of the operational spectrum rather than at the other end, and that the three factors governing this trend are as applicable to the situation in Europe as they are elsewhere. In this connection the increasing presence of Russian ships, especially in the Mediterranean and around the coasts of Africa, is not without significance because valuable help to subversive movements in maritime areas can be provided from the sea.

Although it is not intended to try and predict the exact situations in which Britain might become involved in countering subversion and insurgency, it is none the less necessary to examine in broad terms some of the contexts in which such contingencies could arise. In this respect the position which Britain will hold in relation to the rest of Europe during the second half of the 1970's is naturally relevant. In a publication entitled *Europe's Futures, Europe's Choices,*[1] produced by the Institute for Strategic Studies, the authors suggest six ways in which Europe might develop and then very sensibly point out that the most likely course for events to follow is a seventh one which they are unable to predict. But although the future is so uncertain there are three factors which can be identified, and consideration of which may be of value. The first of these is that whatever does evolve in Europe will take a long time to come about and that so far as the 1970's are concerned, Britain is still going to be concerned with defending her own national interests. Although these interests may be becoming increasingly close to those of other European countries, and although defence ties with other European countries may become stronger, there is little likelihood that any form of European grouping as distinct from NATO will emerge that can identify the interests of its members collectively, or take over the responsibility for defending them collectively, in so short a space of time. The second factor is that, irrespective of the final outcome in Europe, Britain will never be able to retreat into a position of complete isolation so far as defence is concerned. For centuries she has been obliged to enter into combinations to provide for her

[1] *Europe's Futures, Europe's Choices.* Edited by Alastair Buchan, Chatto and Windus, 1969.

safety and the situation remains the same to-day. If Britain becomes more closely integrated into Europe she will perhaps be less directly concerned with America, but at the same time Europe as a whole will be dependent on an understanding with America in one form or another. If Britain does not become more closely involved in Europe she will automatically become increasingly dominated by America, because not only will she require a close association for defence purposes, but she will also become more intimately connected in the economic field, which is another way of saying that her interests will be more fully identifiable with those of the United States. The third factor is very obvious but frequently overlooked; it is that countries are obliged to fight where their interests demand that they should, and this is not necessarily along their geographical frontiers. Thus even if Britain becomes totally submerged in a European community, that community, and therefore Britain, must be prepared to fight wherever its interests require, which might well be outside Europe. Similarly if Britain becomes more closely bound up with the United States she may be obliged to fight where the joint Anglo-American interest is threatened.

The uncertainty of the situation so far ahead as the second half of the 1970's is just as marked in relation to events which are not directly connected with Britain's position in Europe and SEATO, CENTO and NATO could all change their form radically over a period of ten years. But one commitment will inevitably remain which is the obligation for maintaining law and order within the United Kingdom. Recent events in Northern Ireland serve as a timely reminder that this can not be taken for granted and in the historical context it may be of interest to recall that when the regular army was first raised in the seventeenth century, 'Suppression of the Irish' was coupled with 'Defence of the Protestant Religion' as one of the two main reasons for its existence. In practice the fact that the army is so heavily engaged in Ireland now makes it unlikely that it will be involved in exactly this task between 1975 and 1980 because it is reasonable to hope that the present emergency will be resolved within five years. Even so there are other potential trouble spots within the United Kingdom which might involve the army in operations of a sort against political extremists who are prepared to resort to a considerable degree of violence to achieve their ends. It is difficult for the

British with their traditions of stability to imagine disorders arising beyond the powers of the police to handle, but already there are indications that such a situation could arise, and this at a time of apparently unrivalled affluence. It has to be recognized that methods of tying down large numbers of policemen and soldiers have been developed for use against governments which rely on popular support and which can not therefore afford to use the sort of ruthless brutality which a dictatorship could use in order to control the situation in an economic way. If a genuine and serious grievance arose, such as might result from a significant drop in the standard of living, all those who now dissipate their protest over a wide variety of causes might concentrate their efforts and produce a situation which was beyond the power of the police to handle. Should this happen the army would be required to restore the position rapidly. Fumbling at this juncture might have grave consequences even to the extent of undermining confidence in the whole system of government.

Before leaving the question of Britain's position with regard to the use of force during the second half of the 1970's it is necessary to look at one further point. Whereas subversion, insurgency and orthodox war have been known for centuries, a form of military operation has recently been developed which genuinely does break new ground, and that is peace-keeping in the sense in which it is defined in the introduction to this book. This is a totally different activity to that which used to be known as keeping the peace or as duties in aid of the civil power, because both of those tasks were concerned with operating on behalf of a government against people who wanted to upset its authority. In other words keeping the peace and duties in aid of the civil power were polite terms used to describe a mild form of countering subversion. Peace-keeping is different because the peace-keeping force acts on behalf of, and at the invitation of, both sides to a dispute, and it is supposed to prevent violence without having recourse to warlike actions against either of them.

A number of peace-keeping operations have taken place during the past fifteen years ranging from the operation of relatively large formations, such as those deployed by the United Nations in the Congo and in Cyprus, to the use of observer teams in Kashmir, Viet Nam and along the Arab-Israel border. Although most peace-keeping operations have been carried out under the aegis of the United Nations it is worth remembering that Britain established and maintained a force working on exactly these lines in Cyprus between Christmas 1963 and the end of March 1964, and during January and February there was a suggestion that it would be replaced by a NATO peace-keeping force or by a Commonwealth one.[1] The fact that it was ultimately relieved by a UN force should not be taken to mean that in the future all peace-keeping operations will be associated with this body. The nature of the task is such that any nation or group of nations may find itself invited to form or take part in a peace-keeping force, and provided that nation has an interest, however remote, in preventing a flare-up of the contest, it may feel obliged to accept.

Despite the Cyprus experience, the likelihood of Britain operating unilaterally in a peace-keeping role is fairly remote, but it could happen if, for example, a divided community having friendly links with Britain particularly asked for assistance of this sort. The likelihood of Britain operating as part of a NATO or Commonwealth peace-keeping force is rather greater. There are countries in many parts of the world, particularly in Africa, which might require assistance of this nature and Britain is well suited to providing it on account of her experience in Cyprus, her knowledge of many of the countries concerned and because her forces have the strategic mobility which enables them to be deployed and maintained quickly and effectively. When considering the likelihood of Britain being asked to contribute to a UN peace-keeping force a new factor arises in that at times there seems to have been an understanding that permanent members of the Security Council should not be asked to provide soldiers. Dag Hammarskjold undoubtedly held this view but this policy never became accepted as a hard and fast rule for United Nations operations, and it is probably fair to say that it is now out of date. It was certainly waived in Cyprus so far as the British were concerned because of the special situation which prevailed there, and it could easily be waived again if circumstances demanded that it should. Relatively few nations have the inclination or the ability to contribute forces, especially the sort required for logistic and command purposes such as transport, repair and communications units. Furthermore, not only does the establishment of a peace-keeping

[1] R. STEPHENS, *Cyprus, a Place of Arms*, Pall Mall Press, 1966, pp.188–189.

force have to be acceptable to the parties to the dispute, but also the nationalities of the contingents has to be agreed upon as well, which further reduces the number of nations from which the force can be drawn. For all these reasons the fact of Britain being a permanent member of the Security Council is unlikely to disqualify her from taking part even if the custom were once more to be regarded as a guide to United Nations operations. It is felt in some quarters that a future Middle East peace-keeping force would need to include contingents from both Britain and France in order to make it effective and it is even possible that Russia and the United States might operate together under the United Nations flag one day if the disturbance which gave rise to the requirement was on a large enough scale and if the interests of these two countries were sufficiently close as to warrant it. It is of interest to recall that Russia suggested that this should happen at the time of the Suez campaign.

All in all the 1970's may turn out to be as stormy as the 1960's if not more so, but it is none the less virtually impossible to plot the path of the storms. All that can be said with confidence is that notwithstanding the reduction in commitments, Britain in common with the USA and many other countries is unlikely to be able to avoid all of the storms, and in dealing with them one of the most important things to realize is that most countries now regard subversion and insurgency as an integral part of one total war and not as a separate subject. Whether in Europe or overseas, the pattern of conflict is such that it is virtually impossible to imagine an orthodox war taking place without an accompanying campaign of subversion and insurgency, although the reverse is by no means true. It is of course necessary to point out that although in the world at large subversion and insurgency are likely to account for most of the operations which take place in the 1970's, it does not necessarily follow that each country should organize its forces primarily to fight this sort of war. In some areas orthodox conflicts are still possible in the context of the special conditions mentioned earlier, and few countries would yet be prepared to dispense with the means of defending themselves against conventional attack in case the conditions which have produced the present situation change suddenly contrary to expectation. Another important consideration is that the nuclear balance itself is not solely based on weapons of mass destruction, but depends to a certain extent on a combination of nuclear weapons and conventional forces, some of which are totally unsuitable for dealing with insurgents. None the less the essential fact remains that any nation preparing to defend itself in the 1970's must be at least as well prepared to handle subversion and insurgency as to take part in orthodox operations, despite the fact that the requirement is bound to lead to heightened competition for resources which are already far from adequate for meeting the needs of orthodox war.

This competition for resources is a difficult matter to resolve, and to some extent the only satisfactory answer is to allocate more resources over all to defence purposes than is being done at the moment. But there is a lot which can be done with the material already at hand and it is probably fair to say that unless our knowledge of the mechanics of fighting subversion and insurgency is improved, any extra resources allocated will be largely wasted. Peace-keeping is admittedly less important but many of the techniques have much in common with countering subversion, and it is therefore worth examining them so as to ensure that the maximum benefit is gained by those who may be involved in this task from resources and time devoted to preparing the army for counter-subversive warfare. It is to be hoped that this chapter has at least succeeded in showing that both peace-keeping and the fighting of subversion and insurgency are likely to face the army in the 1970's and that preparations for taking part in these operations should be afforded a proper priority in relation to that given to preparations for orthodox war.

Chapter 2

Enemy Aims and Methods

It is sometimes said that insurgents start with nothing but a cause and grow to strength, while the counter-insurgents start with everything but a cause and gradually decline in strength to the point of weakness. Although this proposition may not be entirely true in the light of all past campaigns, no campaign of subversion will make headway unless it is based on a cause with a wide popular appeal. This derives from the fact, already mentioned in relation to the difference between orthodox wars and wars based on subversion, that the instigators of the campaign rely on the people to overthrow the government once they have been properly indoctrinated and organized. As Roger Trinquier says:
  'The Sine Qua Non of victory in modern warfare is the unconditional support of the population'.[1]

The selection of a good cause often poses severe problems to the organizers of subversion because the real reasons for the campaign may not be such as to attract the population at all. This is particularly likely to be the case when one country is trying to stir up trouble in another one, merely to further its own interests. Yet, if no cause exists it will have to be invented. If a genuine one exists but is not capable of attracting sufficient support, it must be amended until it does. If a good one exists but has lost its appeal for one reason or another, it must be revived. If it is absolutely impossible to produce a cause with enough popular appeal, the enterprise will have to be abandoned because it will be found useless to try and promote subversion or insurgency without one. Guevara lost his life and brought disaster to his followers in Bolivia partly because he insisted on basing his campaign there on a cause – revolution throughout South America – which did not command popular support in Bolivia. In fact weaknesses in causes often offer those involved in fighting subversion excellent opportunities for damaging their opponents, providing the situation is correctly appreciated. It is perhaps worth looking at the causes for which the insurgents ostensibly fought in some recent

[1] ROGER TRINQUIER, op. cit., p.7.

campaigns in order to see the extent to which the organizers had to go in order to present their own aims in popular form.

No problems faced the leaders of the National Liberation Front in Algeria because the cause for which they fought, that is to say independence and self-government, could be presented to the people as it stood. Although the communists made an attempt to gain control of the movement for their own purposes, they were effectively thwarted[1] and the final outcome was exactly in accordance with the aims of the campaign. In Cyprus, too, EOKA's aims of independence from Britain and the right to elect union with Greece could be presented without amendment to the Greek Cypriots who formed a large majority of the island's population. In Indo-China the communist party has been absolutely single-minded in its aim since the 1920's: throughout it has sought to gain control of the country so as to impose its own political doctrine on it.[2] But from the start the communists realized that this programme would receive little popular support, so a series of more popular causes have been presented to the population designed to take advantage of prevailing circumstances. First the people were exhorted to turn against the French,[3] then during the war it was the Japanese, and afterwards it was the French again. Always communist interests were advanced under a smoke screen of patriotic nationalism because communism itself had far too narrow an appeal. In the later stages the communist party worked through a national front organization which published a different and more attractive programme than that put forward by the communist party itself.[4] In the Philippines the communist party conducted two campaigns designed to establish itself in power, and on each occasion the cause was one designed to appeal to a wider section of the community than its own supporters because it was recognized that communism had little appeal either for the people in the villages or for the fighters themselves.[5] In the first campaign the communists worked through an organization called the Peoples Army against the Japanese, but they ensured that an appreciable proportion of its effort was devoted to the business of

[1] EDGAR O'BALLANCE, *The Algerian Insurrection*, pp.56–59.
[2] EDGAR O'BALLANCE, *The Indo China War*, Faber, 1964, p.19.
[3] Ibid., p.29.
[4] GEORGE TANHAM, *Communist Revolutionary Warfare*, pp.130–131.
[5] VALERIANO AND BOHANNAN, *Counter Guerilla Operations*, Pall Mall, 1962, p.93.

eliminating Filipinos whom they regarded as a long-term threat to communism.[1] For the second campaign the name changed to the more openly communist one of the People's Liberation Army but the cause was still not presented as raw communism; this time it was thinly disguised as land reform.

One of the most remarkable instances of a cause being manipulated, if not invented, in order to make a wide appeal is afforded by the Mau Mau movement in Kenya. In this case educated African nationalists clearly wanted to get control of the government so as to steer the country towards independence, but they realized that such an idea was far too vague to appeal to the tribally minded people of the time. They therefore decided to concentrate on one relatively minor grievance which existed by reason of the fact that when the country had been settled by Europeans in the first decade of the present century, a very small area of Kikuyu land had been occupied because, at the time, there were no Kikuyu living there.[2] None of the other land settled had ever belonged to this tribe, and the arrangements made with the other tribes concerned had been perfectly satisfactory. It is a measure of the understanding shown by the leaders of the rebellion that they should have selected this issue for their cause because it immediately awoke a response throughout the length and breadth of the tribal area which could never have been matched by any political or economic programme however firmly based on reality.[3] Night and morning prayers were offered up for the recovery of the stolen land whilst those praying held aloft a handful of sacred soil.[4] The Mau Mau gangs were known collectively as the Kenya Land Freedom Army[5] and many of their songs centred around this crucial issue.[6] In the end thousands gave their lives for it, neither knowing nor caring that the original area concerned only extended to a few square miles. In their minds they had come to regard any land occupied by a European as their land, and it is in men's minds that wars of subversion have to be fought and decided.

[1] ARTHUR CAMPBELL, *Guerillas*, Arthur Barker, 1967, pp.124–125.
[2] L. S. B. LEAKEY, *Mau Mau and the Kikuyu*, Methuen & Co., 1952, pp.66–67.
[3] Ibid., pp.105–106.
[4] F. E. KITSON, *Gangs and Counter Gangs*, Barrie and Rockliff, 1960, p.179.
[5] FRED MAJDALANEY, *State of Emergency*, Longmans, 1962, p.157.
[6] F. E. KITSON, op. cit., p.183.

Most of those who study the works of Lenin, Mao Tse Tung and Vo Nguyen Giap like to interpret their teaching in terms of the various stages through which subversion and insurgency has to progress, e.g., organization, terrorism, guerilla warfare, and mobile warfare,[1] and their writings are readily available to those who find such methods of presentation helpful. In practice an understanding of these theories is not essential and can even be misleading because subversive campaigns develop along such widely divergent lines. The really important point is that the leaders of a subversive movement have two separate but closely related jobs to do: they must gain the support of a proportion of the population, and they must impose their will on the government either by military defeat or by unendurable harassment. In terms of time, it is obviously desirable and usually necessary to get some support from the population before embarking on operations, and in most campaigns of recent times the organizers have devoted many years to the task before making any hostile move against the authorities. On the other hand there have been occasions when the two aspects of the business have started at virtually the same moment, and in any case securing and maintaining the people's support has to continue throughout hostilities right up to the end of the campaign.

Subversive movements are particularly vulnerable during the period when the population is being organized to produce support, therefore it is worth looking to see how long such a period may be expected to last. In the Philippines the communist party issued its Strategy Directive in 1946 but it was not until 1950 that the People's Liberation Army started their military offensive, although before that, coercion was used on unco-operative elements of the population and on occasions force had been used defensively against the government.[2] The time spent in preparing the ground in Algeria was much shorter, and in fact the insurrection was launched suddenly and violently only eight months after the political body which sponsored it was brought into being,[3] but it is necessary to point out that previous organizations committed to fighting a war of independence had been in existence for at least

[1] JOHN J. MCCUEN, *The Art of Counter-Revolutionary War*, Faber & Faber, 1966, p.40.

[2] ARTHUR CAMPBELL, op. cit., pp.126–127.

[3] EDGAR O'BALLANCE, *The Algerian Insurrection*, 1967, pp.37–39.

five years.[1] In Cyprus Grivas started preparing as early as 1951 for the insurrection which broke out in 1955, but for the first four years his activities were concerned with planning, building up supplies and dealing with Greek and Cypriot national leaders. He managed to do all that was necessary in the way of organizing armed groups and a network of supporters' cells in the five months which elapsed between November 1954 and March 1955,[2] but the situation in Cyprus was unusual to the extent that the aim of the movement, i.e. union with Greece, was very widely accepted by the Greek Cypriots. From earliest childhood they had been brought up by the church and by the schools to pray and work for ENOSIS so that no indoctrination was needed; only organization. By contrast indoctrination of the population had been going on in Kenya for at least four years before the outbreak of the Mau Mau uprising in 1952,[3] and in Malaya the communists had started preparing for the rebellion which broke out in 1948 as soon as the Japanese had been driven out of the country in 1945,[4] if not earlier.

At this point it is necessary to mention the ideas of Fidel Castro and Che Guevara. Although Mao Tse Tung had altered the classic Leninist revolutionary doctrine to the extent that he had lain emphasis on an action in rural as opposed to urban areas, he none the less retained and expanded on the idea that it was necessary to organize the population as a prelude to armed revolt. Castro and Guevara, however, claimed that their experiences in Cuba showed that it was not necessary to wait for all the revolutionary conditions to be present but that insurrection could itself create these conditions.[5] They developed what has come to be known as the foco theory which is that a small group of armed insurgents operating from a remote part of a country can act as a focus for the various discontented elements in that country and thereby channel all the latent energy available into action for the defeat of the government. In theory the foco, i.e. the small armed group, does not itself defeat the government but merely causes it to be

[1] EDGAR O'BALLANCE, op. cit., pp.34–35.
[2] GEORGE GRIVAS, *The Memoirs of General Grivas*, (ed. Charles Foley), Longmans, 1964, pp.13–25.
[3] FRED MAJDALANEY, op. cit., pp.58–69.
[4] RICHARD CLUTTERBUCK, *The Long, Long War*, Cassell, 1967, p.22.
[5] CHE GUEVARA, *Guerilla Warfare: A Method*, Normount Armament Company, 1966, p.2.

defeated by a combination of all the revolutionary forces concerned including itself.[1] Castro and Guevara also maintained that
although in the insurrectional phase firm leadership is necessary,
it is possible to do without another of the Leninist requirements,
a vanguard party of the working class,[2] which can be formed after
victory is won. According to Castro, the Cuban campaign was
fought with the help of men of all ideas, of all religions, and of all
social classes. It was only after victory had been achieved that
there was a rapprochement between Castro and the communist
party which virtually united the peasants and the working classes
against the rest.[3] But although insurrection without organization
worked in Cuba, it was nearly defeated and it is only necessary to
read Guevara's account of the war[4] to realize how close the enterprise was to failure on a number of occasions during the first six
months. Guevara's subsequent campaign in Bolivia, which was
also based on the foco theory, failed completely in under a year
because the absence of a supporter's organization prevented him
from getting any recruits and made the obtaining of food, medicine
and supplies almost impossible. His Bolivian diary[5] gives a day-
to-day account of the disintegration of his force and is one of the
most instructive documents available. In fact, as a result of the
Bolivian experience it seems unlikely that future operations of
this sort will be launched without some sort of preparation
amongst the population and although the foco theory could perhaps be adapted to suit the needs of various forms of urban subversion the Castro form of insurrection can probably be regarded
as exceptional.

Reverting to more typical cases it would seem normal for the
period of preparation to be measured in years rather than in
months, and this is not surprising when the actual mechanics of
the job are examined. In the first place the organizers have to form
a new party or front to promote their cause unless an existing one
has the necessary popular appeal, which is unlikely. Having
formed a party, it has got to spread so that branches of it exist
throughout the country, and then those entrusted with running

[1] REGIS DEBRAY, *Strategy for Revolution*, Jonathan Cape, 1970, p.39.
[2] Ibid., p.53.
[3] ANDREW SINCLAIR, *Guevara*, Fontana, 1970, p.50.
[4] CHE GUEVARA, *Reminiscences of the Cuban Revolutionary War*, Pelican, 1969.
[5] CHE GUEVARA, *Bolivian Diary*, Jonathan Cape/Lorrimer, 1968.

the branches have to initiate cells in factories, on farms, and in other suitable places throughout their area. These cells in their turn have to work on the people around them in order to get the degree of support required, which may range from benevolent neutrality to the provision of funds, equipment, or intelligence. The means by which branches get hold of people to run the cells, and the way in which these people in turn get what they want from the population as a whole vary considerably but are likely to include the spoken, written and broadcast word with coercion in the form of blackmail and even physical violence thrown in where necessary, providing that it can be done without attracting too much attention from the authorities. All of these activities take time particularly if funds are limited.

People seldom write about their doings in the preparatory stages of subversion, which is unfortunate because a clear understanding of their problems at that time would be of great value. An excellent account of events in this phase is, however, available in a book which deals with the activities of the African Party for the Independence of Guinea and Cape Verde.[1] The party itself was formed in 1956 and immediately started building up support throughout the country. By 1960 the Portuguese had obtained enough information to take action and the secretary-general of the party was obliged to move across the border to Conakry where he set up a headquarters and a training school. In 1962 the movement went onto the offensive with a series of sabotage raids.[2]

The story of a man named Antonio Bana,[3] who became involved in the movement in 1957 or 1958, well illustrates the problems which face those responsible for organizing the population: at that time he was working in a garage in the capital of Portuguese Guinea, and was about twenty years old. Bana first became involved when a friend told him that a party had been created to fight against the Portuguese so that the people could have liberty and a new life. Thereafter he met his friend several times and soon volunteered to join. His first job was to help mobilize the people in the capital for which purpose he arranged small meetings in places he knew with people he could trust, in order to tell them

[1] GERARD CHALIAND, *Armed Struggle in Africa*, Monthly Review, 1969.
[2] Ibid., pp.21–22.
[3] Ibid., pp.72–83.

that the time for action had arrived. Later on the president of the party met him and explained things to him in more detail. Soon afterwards he and some others were selected to go out into the countryside and 'mobilize' the peasants. This probably happened in 1959. The party's teaching with regard to the recommended method for building support in an area was that the first approach should be made to the principal elder in a village. This man was not the administrative leader appointed by the government and known as the chief who usually came from a different region and often from a different tribe. The secretary-general of the party gave Bana and his friends detailed advice on how to approach the village elders by making them play games in which one man took the part of the party worker and the rest pretended to be the village elders.

Bana then describes how things worked out in practice. He started by finding out all he could about the principal elder in a particular village, and if he got encouraging reports he presented himself to the man as a traveller and asked for hospitality. In accordance with traditional practice the elder would then have a meal of rice prepared. If no chicken or meat was provided with the rice, and it seldom was provided, Bana would ask the reason, expecting the stock answer that the man was too poor to provide it. His next move was to enquire why this should be the case with a much respected person at the end of his working life, and he would usually be told that it was the result of taxes imposed by the Portuguese. This gave him an opportunity for talking about the party and for suggesting that the elder should mention it to other people in the village who might be prepared to help the cause. He would finish by inviting the elder to come with a few friends to a meeting to be held outside the village next time he, Bana, came to the area. Meanwhile the party would ensure that some leaflets were distributed in the village which would have the effect of stimulating interest and building up an impression of the party's power. Party broadcasts from Conakry also helped to strengthen its influence.

At the meeting with the elder and his friends, Bana would answer questions about the party and its programme and then suggest that they should pass the news on to other people in the village who they trusted. Later he would call a further meeting of this wider circle and by discussion discover which of them were best suited to becoming party militants. Some of them would be appointed with the agreement of the rest to assume party responsibility in the village. In 1961 Bana went on a course at the party school in Conakry where he met other members, some of whom had just returned from a year's military training at a place which is not mentioned. The course gave him confidence, but when he returned to his area he found that the Portuguese had embarked on action designed to discover and remove those involved in the movement. The Portuguese evidently had some success in discouraging certain people but others merely became more determined. Bana describes the problems of isolating and neutralizing the influence of those who opposed the movement on the grounds that it brought the wrath of the Portuguese down on their heads. He concludes this part of the narrative by saying that mobilizing the population is a much harder thing than armed struggle itself.

1962 started badly for Bana and his friends because Portuguese pressure obliged them to cross the border into Senegal where they were locked up as a result of a twist in local politics. Eventually they got back to their areas and by taking advantage of errors in the Portuguese propaganda programme and of the hatred engendered by their operational methods, they managed to re-establish themselves in the villages. Later in the same year they embarked on a campaign of sabotage but this brought down such a concentration of Portuguese effort that some of the party leaders had to fall back over the border again. However, Bana makes the point that those who, like himself, had managed to get their villages well organized, were able to merge into the population and remain. From this point the campaign is one of insurrection and the account tells of the arrival of weapons and of the start of the fighting proper. The six-year period of preparation was over.

Six years is undoubtedly rather longer than the normal time required for preparation, and gives an indication of the difficulties and frustrations involved in getting a primitive people to organize themselves effectively, as well as being a reflection on the fact that the party had to start from rock bottom in Portuguese Guinea because no preliminary work had been done before it was formed. In many places there is no doubt that the groundwork has been done already, particularly in Europe, where for years the communists have been infiltrating their agents and sympathizers into factories, co-operatives, educational establishments, unions, government departments, the police, the services, and even the church.[1] In most places arrangements have also been made to enable the network to continue to function even if the government were to swoop down and arrest all known communists.[2] In such countries it would not take long to build up a popular front and be ready to launch an insurrection.

It is interesting to speculate whether as much time is required during the organizational period if the projected insurgency is designed to take place in an urban area as it is if it is planned to take place in a rural setting. This is a difficult question to answer because there is nothing comparable to Antonio Bana's story which relates to a purely urban situation. Insurgent activities took place in Jerusalem, Nairobi, Algiers, Nicosia and Saigon but always in conjunction with a rural campaign, so even if less time was required for building up the necessary organization in the cities than in the country surrounding them, the fact would not be apparent because the campaign could not start until the organization had been built up to a sufficient extent throughout the country as a whole. In Aden the urban campaign was virtually separate from the one carried out by dissident tribes in the Protectorate and it therefore affords a better indication of the time required for preparation. In this case the decision to embark on the campaign was taken in October 1963 and the first shots were fired in November 1964,[3] i.e. at least a year was spent in the organizational phase.

A better basis for estimating the time required to build up an urban organization can be obtained by working out exactly what might have to be done in order to build one up. In this connection the ideas of the American writer J. J. McCuen, who is rapidly becoming one of the foremost experts in this field, are relevant. In a military paper which he wrote in 1968 but which he has not yet published, he equates the well-known phases of revolutionary warfare as expounded by Mao Tse Tung with three phases for urban insurgency which he calls organization, civil disorder and terrorism. In the organization phase he envisages the building up of three inter-connected hierarchies: a politico-administrative

---

[1] IAN GREIG, *Assault on the West*, pp.244–251.
[2] Ibid., pp.252–254.
[3] JULIAN PAGET, *Last Post : Aden 1964–67*, p.115.

organization, a control structure, and a functional system of associations, clubs and other groupings designed to carry out specific tasks. Maintaining that a fully developed organization is a necessary prelude to success he describes how a group of would-be insurgents could establish themselves in positions of power within a political organization and then turn it into an instrument of violence, whilst building up a control structure at every level and infiltrating into various clubs and associations. The description leaves little room for supposing that the task could be achieved at an appreciably faster rate than the build up of a subversive organization in a rural area. Naturally, if Castro's foco theory was found to be practicable in an urban setting, this would speed up the process of organization, and at first sight it might appear that small subversive groups could operate effectively without extensively organized support because supplies are easily obtained, and because good cover is afforded by a large built up area. In practice if the government is at all slow in developing a system for identifying the insurgents they will probably survive for long enough to attract the support of a significant proportion of the population, and if this happens the government's task will become immeasurably harder. But if the government builds up a really effective intelligence organization quickly, insurgents operating without the insulation provided by a closely linked system of secure cells, will be eliminated before they can become dangerous, because they will have to make their own contacts with the uncommitted population of the city.

During the early stages of preparation in rural or urban areas the party has little need for armed forces because it is concerned with gathering support by means of the various forms of persuasion and non-violent coercion already described. But eventually the moment arrives when persuasion has to be supplemented by rougher methods and strong-arm groups have then to be brought into being, in order to carry out the policies of the party leaders and committees. To start with these groups may work on a part-time basis and live amongst the population, but once they start to operate in earnest the government is bound to launch its counter measures if it has not already done so. At this stage, providing conditions are suitable, some at least of the strong-arm groups will probably take to the hills or forests to act as the nucleus of larger guerilla parties. Similarly small terrorist and sabotage groups may form in urban areas if the campaign is designed to develop in that direction. A further refinement of the system is that in the later stages of the most highly developed insurgencies elite groups are formed by taking the best fighters out of the guerilla bands and forming them into regular units. The guerillas are then responsible for day to day actions in their own areas, whilst the regulars operate wherever they are required.

The size of the armed groups naturally depends on the plan of campaign which in turn depends on such factors as the cover afforded by the terrain, the proximity or otherwise of a friendly country, the availability of weapons, recruits, food and money, and above all on the political situation. The Greek civil war affords a good example of operations carried out by large insurgent bands. By 1947 the basic unit of the Democratic Army was the battalion 200–250 strong[1] and in 1948 several of these were grouped together into brigades of 1000–2000 strong[2] although in the event, this move which went hand in hand with the adoption of orthodox as opposed to guerilla tactics, turned out to be a serious mistake.[3] In Algeria the basic unit was the battalion of approximately 350 men which was organized into companies and platoons, and although in some cases the whole unit lived and operated together, it was more usual for companies to be allocated to areas of their own.[4] As in Greece, the Algerians tried forming regiments consisting of two or three battalions[5] but, as in Greece, the attempt was a failure and had to be abandoned.[6] In Malaya when the emergency was at its worst the insurgents lived and operated as companies 100 strong[7] but within two or three years they broke up into platoons of 20–30 strong.[8] In Cyprus the largest groups were only 8–10 strong and most of them consisted of 5–6 men.[9]

Tasks allocated to armed groups also depend on the plan of

[1] EDGAR O'BALLANCE, *The Greek Civil War*, Faber and Faber, 1966, p.151.
[2] Ibid., p.182.
[3] Ibid., p.212.
[4] EDGAR O'BALLANCE, *The Algerian Insurrection*, pp.72–74.
[5] Ibid., p.117.
[6] Ibid., p.122.
[7] RICHARD CLUTTERBUCK, op. cit., p.46.
[8] Ibid., p.86.
[9] GRIVAS, *Guerilla Warfare*, p.33.

campaign and may include violent coercion of elements in the population opposed to the party, attacks on weak government forces such as Home Guard or Militia units whose job it is to protect loyal sections of the population, and ambushes or attacks on small bodies of the government's forces to the extent necessary for keeping the campaign in the public eye as part of the business of harnessing world opinion or for the purpose of getting hold of government weapons and equipment. Tasks may also include acts of sabotage and terrorism designed to ensure that the government deploys disproportionately large bodies of its own forces on protection duties and searches, and carefully calculated acts of revolting brutality designed to bring excessive government retaliation on to the population thereby turning them against the government. Eventually, if circumstances are favourable, armed groups may be required to challenge the forces of the government in open warfare. But despite the teaching of Mao Tse Tung and his disciples, action of this sort often back-fires as it did in Greece and Algeria. It would even be possible to argue that in Mao's own campaign against Chiang Kai-Shek it was not the operation of his regular forces which brought about the final victory, but the cumulative harassing effect of guerilla warfare combined with the intrinsic rottenness of Chiang's political position. In urban areas armed insurgents may have the additional task of fomenting demonstrations and riots. These have the dual purpose of using up large numbers of government forces and of establishing a further link between the uncommitted members of the population and the insurgent organization. In order to do this the insurgents may well be obliged to exploit local grievances which are in no way connected with the aims of the movement and which may vary from place to place. This poses no problems for the insurgents but tends to confuse the forces of the government who may not initially recognize the directing hand of the subversive organization.

The relationships which have existed between parties and their armed groups in the past have varied considerably between one campaign and another, although certain essential features are common to most of them. Broadly speaking it is almost always recognized that the political leadership should take precedence over the military because the ultimate aim is usually political, and the means of achieving it are also political in so far as they are concerned with gaining control of the population. The important and obvious differences which have arisen have done so as a result of the differing circumstances and personalities concerned. For example, to take an extreme case of divergence from the normal, Castro in Cuba and Guevara in Bolivia refused to acknowledge the supremacy of the official communist parties in the countries concerned because their intention was to create new revolutionary parties centred round their own focos after victory. They therefore formulated theories to justify their actions such as those expounded later by Regis Debray when he said –

'No political front which is basically a deliberate body can assume leadership of a people's war; only a technically capable executive group, centralized and united on the basis of identical class interest can do so; in brief, only a revolutionary general staff'.[1]

In another passage he expanded on this by saying –

'Under certain conditions the political and the military are not separate but form an organic whole, consisting of the people's army whose nucleus is the guerilla army. The vanguard party can exist in the form of the guerilla foco itself. The guerilla foco is the party in embryo'.[2]

On other occasions, as with the Mau Mau in Kenya, the militant arm started by being fully subordinated to the political wing of the movement but eventually took over control in some areas as the political organization was broken up by government action. There would seem to be little value in working through a whole series of examples from the past in order to illustrate these facts, but it is worth describing one typical system in order to demonstrate the essentials of the business.

As good an illustration as any is afforded by the command system set up by the insurgents in Malaya.[3] At the top of the tree was the central committee of the Malayan Communist Party which operated under the chairmanship of the secretary-general, and which included on it the commander of the Malayan Races Liberation Army. Under the central committee were committees for each of the states or regions of Malaya. Each State Committee consisted of a state committee secretary and a number of members

---

[1] REGIS DEBRAY, *Revolution in the Revolution?*, Pelican Books, 1968, p.85.
[2] Ibid., p.105.
[3] RICHARD CLUTTERBUCK, *The Long Long War*, p.88.

each of whom was the secretary of his own district committee, that is to say the next committee down. Each district committee controlled the activities of a number of branch committees each consisting of several members. At every level the committee was supported by a number of guards, escorts and individuals concerned with the reproduction of propaganda. From the start of the emergency the central, state and district committees all operated from camps within the jungle but many of the branch committees lived with squatters on the jungle edge. Later these committees also moved inside the jungle, but their job remained unaltered which was to maintain contact with, and control of, the cells which had been set up within the villages. The Malayan Races Liberation Army was split into companies and later into platoons as described earlier. Although in theory the commanders of these units had a chain of command upwards to the commander of the army for organizational matters, they received their operational orders from the committee secretary in whose area they were operating. For example, if a platoon which was working in a district was required to do a particular job, it would receive orders from the district committee secretary to get in touch with the branch committee in whose area the job was to be done and carry out the work required of it.[1] The system is shown diagrammatically in Figure 1.

Most of the campaigns examined so far have been genuine home-grown products, that is to say they were directed and carried out by citizens of the country concerned for what they regarded as the benefit of their countrymen. In many cases other nations took advantage of these uprisings and gave support of one sort or another, either because they genuinely sympathized with the aims of the subversive movement, or because it was in their interest to see the existing government replaced by the insurgents. It is now necessary to look briefly at two variations to the basic pattern, the first of which is the subversive campaign deliberately fostered by an outside country for its own purposes.

By reason of the nature of subversive campaigns, an outside country can only foster and direct one in another country if it can find a cause which will appeal to the people of that country, and which does not run counter to its own aims. This is not as easy as it sounds. Having found a cause it is then necessary to go

[1] RICHARD CLUTTERBUCK, op. cit., p.91.

FIGURE 1

ORGANIZATION OF MALAYAN COMMUNIST PARTY (MCP)

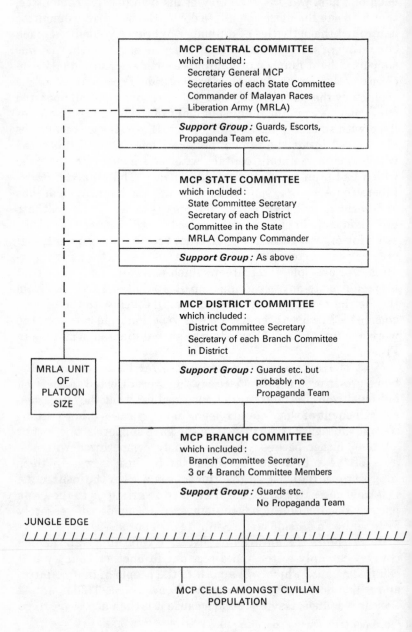

MCP CENTRAL COMMITTEE
which included:
  Secretary General MCP
  Secretaries of each State Committee
  Commander of Malayan Races
  Liberation Army (MRLA)

Support Group: Guards, Escorts,
Propaganda Team etc.

MCP STATE COMMITTEE
which included:
  State Committee Secretary
  Secretary of each District
  Committee in the State
  MRLA Company Commander

Support Group: As above

MCP DISTRICT COMMITTEE
which included:
  District Committee Secretary
  Secretary of each Branch Committee
  in District

Support Group: Guards etc. but
                              probably no
                              Propaganda Team

MRLA UNIT
OF
PLATOON
SIZE

MCP BRANCH COMMITTEE
which included:
  Branch Committee Secretary
  3 or 4 Branch Committee Members

Support Group: Guards etc.
                              No Propaganda Team

JUNGLE EDGE
///

MCP CELLS AMONGST CIVILIAN
POPULATION

through the business of building up a political front, together with armed forces if required, before the campaign can be got going, all of which takes time. Certainly examples of this sort of insurrection can be found such as the one fostered by North Viet Nam in South Viet Nam, but this might be regarded as a special case because of the fact that the two countries had only recently been established as separate nations. The fact is that the deliberate instigation of subversion by one country in another, except in conjunction with large-scale orthodox operations such as those which took place in the two World Wars, can only succeed after a long period of preparation. It is none the less necessary to remember that a great deal of preparation has been carried out by communist countries in non-communist ones, especially in Europe. Where a campaign is run on these lines it is likely that the outside country will restrict its contribution in terms of manpower to the minimum so that it does not detract from the national appeal of the insurgency. A few specialists to carry out technical work on the spot may be an asset because they serve as a living symbol of outside support, but they will certainly not want to give the impression of directing the campaign. In fact most communist countries are known to object in principle even to the attachment of specialists to insurgent forces if it can be avoided, because of the adverse effect it might have on opinion both inside and outside the country concerned. They greatly prefer the system whereby potential insurgents come to the helping country for training and indoctrination, and then return to spread their knowledge and beliefs.

The second variation which merits examination is one in which a country fosters subversion in another country, and sets it off to coincide with an orthodox military assault or with the imminent threat of one. It is not easy to find good examples of orthodox military aggression being carried out in conjunction with large-scale subversion in the enemy's own country. Both World Wars provide examples of subversion being used in friendly occupied countries, in conjunction with campaigns of liberation, but this provides no proper parallel. A better illustration is afforded by the way in which the Germans used sympathizers in Czechoslovakia to carry out acts of sabotage and to secure landing grounds in concert with the advance of their main army,[1] and by

---

[1] OTTO HEILBRUNN, *Warfare in the Enemy's Rear*, Allen and Unwin, 1963, p.26.

the way in which they activated partisans in the Caucasus in order to divert Soviet forces in the Russian campaign.[1] But the Germans never had sympathizers in the countries which they attacked, organized to anything like the extent to which communists are organized in non-communist countries today. In order to illustrate the point properly it would be necessary to describe an orthodox assault by a communist country on to a non-communist one, but with the altogether untypical exceptions of the Korean War and of the Indian frontier dispute no such example exists, for the simple reason that no communist country has yet attacked a non-communist one, except by means of a subversive campaign. All the same there is always a first time, and if it is considered necessary to be able to resist an attack by orthodox forces from such a quarter, it is at least as necessary to be able to resist one which takes place in conjunction with a subversive campaign. The threat was foreseen over fifteen years ago by Otto Heilbrunn, who wrote:

'But there is no doubt that in any future war in which the USSR or Red China are involved, the Communist Party in the opposing countries will organize guerilla movements for the fight against the armies of their own countries. . . . We must wake up to this threat of supranational communism in our own rear. We have been warned as we shall see by Stalin himself'.[2]

There are several different ways in which a threat of this sort could develop. For example the subversion might come first, to be followed up by an orthodox operation which would be presented to the world as intervention at the request of the insurgents representing the majority of the people. Alternatively the threat of an orthodox assault might come first. In this case tension would be built up in such a way that an assault appeared to be imminent, so that the forces of the opposing country were drawn into their mobilization positions behind the frontier, at which moment strikes, riots and acts of sabotage might break out which were far beyond the capacity of the police to handle. While government forces were held along the frontier by the threat of a military advance, insurgency on a significant scale could develop which,

[1] Ibid., pp.38–39.
[2] DIXON and HEILBRUNN, *Communist Guerilla Warfare*, Allen and Unwin, 1954, p.xv.

amongst other things, might go some way towards paralysing the army along the frontier. Unless the country had enough uncommitted troops properly trained and equipped to fight the insurgents, it would have to sue for a settlement which would probably involve accepting a government formed by the insurgents themselves. Alternatively the country could move its forces away from the frontier to fight the insurgents which would leave the frontier open and enable the aggressor nation to overrun the country and establish the insurgents in power by force. Obviously moves for both sides in a situation of this sort can only be envisaged in the context of the environment as a whole which would have to include the disposition of conventional forces, the availability of reserves, and the position regarding the use of nuclear weapons if available. Where, as in Europe, nuclear weapons and conventional forces are jointly designed to act as a deterrent, the ability to deal with the insurgent threat should also be regarded as part of the deterrent because if the aggressor is able to spark off an insurgency which can partially or wholly neutralize his opponent's conventional contribution to the deterrent, he will have gone some way towards discrediting his deterrent as a whole which could be exceedingly dangerous.

Fortunately the enemy has some difficult problems to contend with in co-ordinating subversion with his other moves in the way described, the principal one being the balance required between getting results quickly on the one hand, and not alienating the population too seriously on the other. For example, if Russia wanted to implement a campaign against the Western Powers in which actual subversion was to be combined with the imminent threat of invasion, it is reasonable to assume that she could immediately create grave difficulties by getting local communists to foment disturbances and commit acts of sabotage. But this would discredit the local communist parties concerned because they would be regarded as traitors to their country in its hour of need and it would later be difficult for Russia to use them in a Quisling type government. In order to get round this handicap the Russians could perhaps build up a National Front in the target country, ostensibly unconnected with the communist party, and this Front could carry out the harassment and sabotage required and then disband leaving the way clear for the official communists to form a government designed to get the best terms possible from the Russians. The system of using separate organizations to operate at different stages of the campaign is not new and was in fact used by the Allies in the Second World War.[1] Although a system of this sort can only be used if there is plenty of time for preparation the danger undoubtedly exists in areas where confrontation between opposing blocs has been going on for many years. It should be recognized as a distinct form of subversion which could easily be used and which the forces of the Western Powers should be capable of handling.

In summarizing the main points made in this chapter it is only necessary to mention the essentials of the basic pattern of subversion and insurgency as seen from the enemy's point of view. In this connection the most important factor is that the immediate object of those organizing subversion is to gain control of the population, and that the normal system for doing this is to select a cause and then form a party which can project it into the population by the organization of a chain of branches and cells, using persuasion and coercion for the purpose. Once support from the population is forthcoming, offensive operations can be started against the government designed to achieve the purposes of the campaign. The operations themselves may be intended to defeat the government by force, but are more likely to be aimed at bringing about a surrender by producing a situation in which it can no longer function. Although the first of these two processes will usually be started before the second, they will both run concurrently thereafter, because, while the engineers of subversion are taking action against the forces of the government, the government itself will be trying to retain or regain the support of the population. It is this interplay of operations designed by both sides to secure the support of the population and at the same time to damage their opponents, that constitutes a subversive campaign.

[1] F. O. MIKSCHE, *Secret Forces*, Faber and Faber, 1950, pp.76–77.

Chapter 3

Civil Military Relations

In attempting to counter subversion it is necessary to take account of three separate elements. The first two constitute the target proper, that is to say the Party or Front and its cells and committees on the one hand, and the armed groups who are supporting them and being supported by them on the other. They may be said to constitute the head and body of a fish. The third element is the population and this represents the water in which the fish swims. Fish vary from place to place in accordance with the sort of water in which they are designed to live, and the same can be said of subversive organizations. If a fish has got to be destroyed it can be attacked directly by rod or net, providing it is in the sort of position which gives these methods a chance of success. But if rod and net cannot succeed by themselves it may be necessary to do something to the water which will force the fish into a position where it can be caught. Conceivably it might be necessary to kill the fish by polluting the water, but this is unlikely to be a desirable course of action.

The fish and water analogy comes of course from Mao Tse Tung, but it has been bent and extended to illustrate the required point. The illustration should not be carried too far because it has important limitations, one of the main ones being that in real life those practising subversion are themselves capable of and indeed much concerned in manipulating the environment, whereas the fish can not do much about the water in which it swims. But from the point of view of those countering subversion the analogy is a good one because it shows how operations, i.e. the rod and net technique, have to be tied in with wider administrative measures, i.e. dealing with the water, in order to kill fish. Translated into normal terms the aim of the government is to regain if necessary and then retain the allegiance of the population, and for this purpose it must eliminate those involved in subversion. But in order to eliminate the subversive party and its unarmed and armed supporters, it must gain control of the population. Thus, in the same way that the first aim of those involved in subversion is to gain control of the people so that the purpose of the uprising can be achieved, so also the first aim of these involved in counter subversion is to gain control of the people because in most cases this is a necessary prelude to destroying the enemy's forces, and in any case it is the ultimate reason for doing so.

Before considering ways in which the government can coordinate its efforts in order to achieve its aim, it is necessary to point out one fundamental matter, which is that few individuals can possibly support a government which is obviously going to lose, even if they sympathize with its policies and detest those of the insurgents. If the government is to be successful therefore, it must base its campaign on a determination to destroy the subversive movement utterly, and it must make this fact plain to its people. If it intends in the long term to relinquish control of the country to another government, it must make plain the fact that it will only do so when that government is strong enough to ensure that the enemy can not gain power. The British campaign in Aden affords a good example of the effects of ignoring this factor. In February 1966 Britain made public her intention of withdrawing her forces from Aden when that country became independent in 1968. In effect this meant that Britain would pull out in 1968 regardless of whether the insurgents had been defeated, and regardless of whether the intended successor government was able to handle them. That at any rate is how the people of Aden saw the situation and few of them doubted that the insurgents would ultimately gain control of the country. As Julian Paget puts it:

'The announcement was a disastrous move from the point of view of the Security Forces, for it meant that from then onwards they inevitably lost all hope of any local support.'[1]

Assuming therefore that the government is at least prepared to avoid undermining its own position completely, the next problem is for it to work out an overall programme designed to achieve its aim of regaining and retaining the allegiance of the population. Such a programme should include measures designed to maintain and if possible increase the prosperity of the country, as well as measures aimed at the destruction of the subversive organization, because not only is prosperity itself a potent weapon in the struggle against those who wish to overthrow the existing order, but also there would be little point in defeating the insurgents

[1] JULIAN PAGET, *Last Post : Aden 1964–67*, p.159.

only to be left with a ruined community. The programme should also cater for rectifying genuine grievances, especially those which the enemy are exploiting as part of their cause, and for attracting support by implementing popular projects and reforms. In this connection it is worth noticing that if all other things are equal, people will prefer to back a limited advance offered by the government than more far-reaching reforms offered by the insurgents because of the greater likelihood of getting something. Finally the normal problems of government unconnected with the uprising will still have to be faced and the measures required for dealing with them must be tied into the rest of the programme if the resources and efforts of the country as a whole are to be used to the best advantage.

It would be relatively easy to analyse a number of past campaigns in order to illustrate the way in which governments have built up programmes containing measures of all these different kinds. Such an analysis would show that on occasions these programmes were put together skilfully, good use being made of all available resources, as demonstrated by Ramon Magsaysay in the Philippines.[1] It would also show that in other places little conscious planning had been done at all and that government policy had merely built itself up as the result of a series of random and unco-ordinated decisions made by different members of the government in the spheres for which they were responsible. It is probable that in most cases it would show that accident and planning had each played their part, the main influence being the situation existing in the early stages of the trouble which was manipulated and adapted in accordance with the needs of the moment. That at any rate is how most government activities are conducted. But the important aspect from the point of view of the military officer engaged in countering insurgency is not to know how to build up overall government programmes so much as to understand how totally interdependent all the various measures must be, and how important it is that they should not cut across each other.

This consideration applies at every level from the top downwards. At the higher echelons of control it is easy to demonstrate because the necessity for close co-ordination between the civil

[1] ARTHUR CAMPBELL, op. cit., pp.129–133.

and the operational effort is apparent to everyone. At the lower levels the need is just as great. On the one hand military officers are required to initiate proposals for wearing down and defeating insurgents which representatives of other government departments have to scrutinize in order to ensure that they do not cut across long-term government aims. On the other hand military officers themselves have to vet action proposed by other departments in pursuance of the government's long-term aims in order to ensure that it is not harmful to the operational effort. Even in the operational sphere civil and military measures are inextricably intertwined. For example the processes involved in dealing with a small band of guerillas in a rural area might include mounting patrols, siting ambushes, resettlement of outlying families in defended villages, imposing curfews on unco-operative sections of the community, searching labourers on their way to work to ensure that they are not carrying food or supplies for the enemy, and the restriction of agriculture in certain areas where it might benefit the guerillas. Soldiers, policemen, local volunteers and junior officials of departments such as the Ministry of Agriculture and the Ministry of Public Buildings and Works would all have to work together on such an operation. There can be no question of the army ever working in isolation except at the very lowest level, and even then it is possible that operations will have to be tied in with the police or home guard.

The United States is well ahead of Britain in its thinking on the overall direction of counter insurgency and counter-subversive operations. A good indication of the way in which the US Army looks at the problem can be got from an examination of the terminology which they now use to describe the various functions involved. In order to stress the importance of positive action designed to gain and retain the allegiance of the population as opposed to purely offensive operations aimed at breaking up insurgent groups, it has stopped referring to 'Counter' insurgency and 'Counter' subversion and redesignated the business as 'Internal Defence and Development'.[1] That part of 'Internal Defence and Development' provided by the armed forces to maintain, restore or establish a climate of order is known as 'Stability Operations', a term designed to emphasize the fact that the

[1] In some places it is known as 'Internal Defence and Internal Development' which is abbreviated to IDID.

purpose of destroying the insurgents is to provide the stability which the country requires so that it can progress and develop. 'Stability Operations' themselves are recognized as being as much concerned with organizing the population as they are with fighting battles. They are broken down into such functions as 'Advisory Assistance', i.e. the task of advising indigenous military commanders in training armed paramilitary and irregular forces: 'Civil Affairs', i.e. the establishment of a co-operative relationship between military units and the population: 'Populace and Resources Control', i.e. action taken by the government to control the population and prevent it from helping insurgents: and 'Psychological Operations' and 'Intelligence', the functions of which are self evident.

Regardless of the terminology used, the process of tying civil and military measures together into a single effective policy is clearly a complicated one. No matter how well aware of the problem the authorities are, they will only be able to solve it if they can devise machinery at every level which can assess all the factors whether they be operational or administrative, short term or long term, make a plan, and then put it into effect. John McCuen puts his finger on the essential issue very well when he says:

'Unified planning, centralized control and a single point of responsibility are the very minimum requirements for a unity of effort which will offer success against a unified revolutionary movement. . . . Unity of effort is however extremely difficult to achieve because it represents the fusion of civil and military functions to fight battles which have primarily political objectives. . . . All the political, economic, psychological and military means must be marshalled as weapons under centralized co-ordination and direction. Unity of effort can be achieved by a single commander as the French advocate. Unity of effort can be achieved by a committee under civilian leadership as the British advocate. . . . '[1]

The rest of this chapter is taken up with showing how this minimum requirement of unified planning, centralized control, and a single point of responsibility can be met under a range of increasingly complicated situations. In the past machinery for tying the military and civil effort together has usually taken the form

[1] JOHN MCCUEN, *The Art of Counter-Revolutionary War*, pp.72–73.

of one of the two systems mentioned above and they are therefore used as the starting point for further explanation. In theory it might seem unnecessarily complicated to consider the problem in terms of adapting two such apparently differing systems but it will be seen that so far as the underlying principles are concerned, there is not such a wide divergence between them. Indeed, although it might be supposed that the Committee system with the representative of the civil government in the chair is superior to the single commander system in view of the overriding importance of gaining the support of the civilian population, it can be demonstrated that both have worked equally well in the past. In any case, so far as future operations are concerned, military officers are not likely to be able to choose between them and will have to operate whichever system the government decides to set up: they should therefore understand both. When considering the more complex situations described in the latter half of the chapter it is necessary to bear in mind that they are developed solely in order to demonstrate the problems of tying together civil and military action, and are not in any way designed as predictions of what may or may not happen in any particular part of the world in the future. That was covered in Chapter 1.

The first of the basic patterns to be considered is the Committee system which has been favoured by the British in recent years. Under this system a committee would be formed at every level on which would sit the senior military and police officers for the area under the chairmanship of the head administrator. The committee might include or co-opt members representing other interests, and it would be served by officials, the most important of whom would probably be an officer of the intelligence organization. Decisions would be taken jointly and implemented by the members using the organizations or units under their command. Each member of the committee would have a superior in his own service to whom he would in any case be responsible for organizational and administrative matters, and to whom he could turn should he wish to protest about a decision made against his advice by the committee of which he was a member. His single service superior could even take up the cudgels on his behalf at the next superior committee if he considered it desirable. In fact the single service superior might well be the representative of that service on the next superior committee anyway. Frequently the system is complicated by the fact that police, administrative and military boundaries fail to coincide, or because in one of the services there is an extra level of command to those used by the other two. For example the military member of the Provincial Committee in the diagram (Figure 2) might be the Brigade Commander which would leave his Battalion Commander without a seat on any committee. But these complications are easily adjusted providing that the individuals concerned understand the system.

Under the single commander system the chairman of the committee becomes a commander and the other members of it become his advisers or staff officers. Usually the commander is a military officer although this need not necessarily be the case. Whether the other two members turn into advisers or staff officers depends on whether they are also commanders in their own right or not. For example in a province where a military officer was the overall commander, his right-hand man with regard to police matters might be the officer in charge of the police in that province in which case he would be an adviser, or he might be a police staff officer in the overall commander's headquarters having no responsibility for running the police in the province, but being solely concerned with his duties towards the overall commander. The two systems are shown diagrammatically as Figure 2.

It is not possible to leave the problem of tying together civil and military action without drawing attention to the arrangements needed at national level in order that one of the systems illustrated above can work. One of the more remarkable manifestations of Vietnamese ineptitude in the early 1960's was afforded by the fact that whereas a form of the single commander system was employed outside Saigon, in the capital itself no effective co-ordination took place, and ministries dealt direct with their own men in the field even on matters of overall national policy.[1] In theory it is not in the least difficult to provide suitable machinery at the top. All that is needed is some form of supreme body which can formulate major policy regarding the conduct of the campaign and pass it on direct to subordinate commanders or committees according to the system in use. This supreme body should consist of the head of the government together with the individuals controlling the most important departments of state such as Finance, Home

[1] ROBERT THOMPSON, *Defeating Communist Insurgency*, p.61.

FIGURE 2

COMPARISON OF COMMITTEE SYSTEM WITH
SINGLE COMMANDER SYSTEM

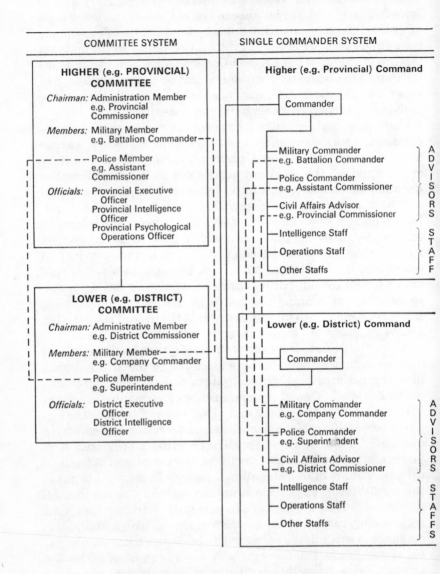

COMMITTEE SYSTEM	SINGLE COMMANDER SYSTEM

HIGHER (e.g. PROVINCIAL) COMMITTEE

Chairman: Administration Member
e.g. Provincial Commissioner

Members: Military Member
e.g. Battalion Commander

Police Member
e.g. Assistant Commissioner

Officials: Provincial Executive Officer
Provincial Intelligence Officer
Provincial Psychological Operations Officer

LOWER (e.g. DISTRICT) COMMITTEE

Chairman: Administrative Member
e.g. District Commissioner

Members: Military Member
e.g. Company Commander

Police Member
e.g. Superintendent

Officials: District Executive Officer
District Intelligence Officer

Higher (e.g. Provincial) Command

Commander

Military Commander
e.g. Battalion Commander

Police Commander
e.g. Assistant Commissioner

Civil Affairs Advisor
e.g. Provincial Commissioner

Intelligence Staff

Operations Staff

Other Staffs

ADVISORS

STAFF

Lower (e.g. District) Command

Commander

Military Commander
e.g. Company Commander

Police Commander
e.g. Superintendent

Civil Affairs Advisor
e.g. District Commissioner

Intelligence Staff

Operations Staff

Other Staffs

ADVISORS

STAFFS

———— Policy link concerning overall direction of the struggle
– – – – Single Service link for organizational matters

Affairs, and Defence. It must be advised by representatives of other relevant agencies such as those dealing with intelligence and propaganda. In campaigns conducted by the British an overall co-ordinator of all military and civil activities has usually been appointed known either as the Director of Operations or as the Security Commander.[1] This man may well hold other appointments and is indeed likely to be the General Officer commanding the troops at the least. He may even be the head of the government as was the case when General Templer in Malaya and Field Marshal Harding in Cyprus carried out this task. But regardless of his other functions his purpose as Director of Operations is to co-ordinate rather than to command.[2] He is in effect the Executive Officer of the Supreme national body.

The problem of co-ordinating civil and military measures is complicated enough when the campaign is the affair of a single nation, but it becomes vastly more so if allies become involved. But once again the guiding principle for tying in the activities of the two countries can be deduced from the ultimate aim of the host nation which is to retain and regain the allegiance of its population. If this is borne in mind, it at once becomes evident that the way in which the ally's help is delivered is as important as the help itself, the main thing being that the host nation should be seen to be at the centre of the picture with the ally coming to its assistance. This impression can only be achieved if the ally is prepared to subordinate itself at every level to the host country. If there is the slightest indication of the ally taking the lead, the insurgents will have the opportunity to say that the government has betrayed the people to an outside power, and that they, the insurgents, are the only true representatives of the nation. In fact they will probably say this in any case, but they will be widely believed if the people see foreigners around the country exercising command over their own troops or worse still laying down policy regarding the civil administration.

In considering the actual mechanics of co-ordinating the activities of an ally with those of a host country, an important factor concerns the sort of help which the ally is proposing to make available. The situation which existed in Greece between 1947–49

[1] JULIAN PAGET, *Last Post : Aden 1964–67*, p.128.
[2] ROBERT THOMPSON, *Defeating Communist Insurgency*, pp.82–83.

is of some interest because at that time both Britain and America were helping the Greeks, the British being responsible for training the Greek army whilst the Americans had the much more important and exacting task of supplying, equipping and advising it on operational matters. In this case the system adopted for co-ordinating policy at the top was for the head of the British Military Mission and the head of the US Military Advisory and Planning Group to have advisory seats on the supreme Greek body, the National Defence Council.[1] At the next level down a Joint Greek – US. Army Staff was formed responsible for planning, equipping and supplying all operations. American officers were also employed as observers down to Divisional headquarters so that they could ensure that plans formulated at the higher level were in fact adopted.[2] The situation in Viet Nam during the early stages of the war there was less satisfactory for two reasons. In the first place neither the Americans nor the Vietnamese had bodies capable of co-ordinating all aspects of their own war efforts, so every different type of aid had to be negotiated between the head of the relevant US agency with his Vietnamese opposite number. More important still was the fact that no supreme council existed for the overall prosecution of the war on which the Americans could be represented. After 1967 considerable improvements were put in hand,[3] further reference to which is made below.

In the light of the Greek and Viet Nam experience, and with the principle firmly in mind of the ally always taking second place to the host country, it is possible to point to three conditions which are likely to govern the successful integration of an ally with a host country in a counter-insurgency campaign. The first of these is that no arrangement will work unless the host country itself has a properly ordered system for prosecuting the war, which must at least include a supreme council and either a committee system or a single commander system at the lower levels, or some recognizable variation of one of them. If such a system does not exist, the ally should bring pressure to bear on the host nation to produce one before agreeing to assist. The second factor is that the ally should be in a position to co-ordinate all its aid in such a way

---

[1] EDGAR O'BALLANCE, *The Greek Civil War*, p.167.

[2] Ibid., p.157.

[3] ROBERT THOMPSON, *No Exit from Vietnam*, Chatto and Windus, 1969, pp.156–157.

that one individual person can represent it on the host nation's supreme council thereby ensuring that the ally takes a proper part in formulating the overall policy for the prosecution of the campaign. This person should of course be the ambassador unless the assistance is entirely restricted to military aid in which case he could be the head of the military mission. The third condition is that the ally should be represented at every level on the host country's committee or staffs according to the system in force, but always in a subordinate or advisory capacity. In this way the ally can feed in military, police or civil aid or advice of any sort as required without having it misapplied, wasted, or used in a way which is not relevant to the successful prosecution of the struggle.

At this point is is worth looking at the arrangement which the Americans make for bringing help to their allies. Although the local situation will seldom allow the concept to be carried out in its pure form, their system is both unique and comprehensive and provides a sound guide for the man on the ground. It is firmly based on the three principles mentioned above. The first thing which should happen when a country indicates that it needs help is that the principal American officials resident in the country, i.e. the ambassador and his leading civil and military advisers, consult with their host country counterparts to determine the particular need and assist in developing a plan. If agreement is reached, the ambassador and his colleagues, who are known collectively as the 'Country Team', submit the requirement to Washington for approval. So far as the Army is concerned, the first step is usually to provide advisers, or 'advisory assistance' as it is called. Trained advisers in various aspects of Stability Operations, such as those mentioned on page 53, i.e. Civil Affairs, Populace and Resources Control, Psychological Operations, Tactical Operations, and Intelligence, are available. Also available are soldiers with such special technical skills as engineering, communications, ordnance, logistics, medical, and military police. Often these advisers are assigned to military advisory groups or missions permanently stationed in the threatened country and are associated with important officials in the indigenous military establishment possibly down to District or Province level and in field force formation headquarters. However, in cases where the requirement is not continuous, it is more economical to assign them to Special Action Forces, which are organized around Special Forces Groups. In these latter cases, the country requirement can be satisfied by forming the advisers into mobile training teams which stay in the country only as long as necessary. For example, in one country the demand may be for a mobile training team of tactical and intelligence specialists and a mobile training team of communications specialists. In another country, the requirement may be for a mobile training team of engineers to advise on civic action. It should be added that only where particular advisers are not available in the Military Advisory Groups or Special Action Forces are they drawn from the US Army as a whole, and this happens primarily with technical specialists.

As a further step, United States field force units can be committed if required into the control framework which has already been established, such as the Military Advisory Group, or under a new tactical headquarters. In either case, the senior military officer is a member of the Country Team. These units can range from small Special Forces teams designed to organize counter-insurgency operations in rural villages, to complete formations of conventional units.

The key to the business lies in the fact that the plan is developed at the top level by the host country in close co-operation with the Country Team and embraces the whole range of civil and military activity. From the military point of view, an important feature of the system is that the advisers, regardless of their Stability Operation specialization, are almost always trained together so they all understand the system as a whole, as well as their own place in it.

A further complication which might occur in the future is that a multi-national organization such as the UN might be called upon to provide assistance in a counter-insurgency situation. The matter would be further complicated if it was decided to back the force with a heavy programme of financial and economic aid. In theory these problems could be resolved without too much difficulty if the three principles previously mentioned were observed. In this case it would be necessary to appoint a political personality to the force who would take the place of the national Ambassador and be the overall co-ordinator of all the assistance provided. He would also have to sit on the supreme council of the host nation and take part in formulating the general policy for the conduct of the campaign. But although financial and economic aid might be channelled in effectively under these arrangements, and although the force itself might arrive and deploy in a satisfactory way, it is unlikely that the actual operations which it subsequently carried out would be of much value unless the various contingents had some common understanding of the military problems concerned in fighting insurgents.

Finally the question arises of the framework which is required for the successful conduct of counter-insurgency operations which take place in conjunction with an orthodox war or with the imminent threat of one. In this case, if the orthodox war or threat clearly concerns aggression into one particular country, then the responsibility for handling subversion or insurgency will obviously lie with that country which must set up machinery to deal with it in accordance with the patterns mentioned earlier. Admittedly new factors will be involved in the formulation of policy and in the conduct of operations arising out of the particular circumstances as, for example, the allocations of forces between the front line and the rear areas, or the need to take more drastic action against the population in order to get results more quickly than would be desirable if there was no external threat. But these are matters which the suggested machinery is capable of handling. Similarly, if one or more allies are brought into the picture, the system suggested earlier for incorporating their efforts can be brought into play, although the ideal solution would be to deploy them in the conventional role thereby freeing more of the Host Nation's troops to fight the insurgents. This course would only be possible if the ally was equipped to take part in the conventional battle, and might not even be possible then, in the light of political considerations.

A more difficult problem presents itself if the actual or potential aggression is aimed at a number of countries as opposed to being aimed at a single one and if disturbances were to occur simultaneously in a number of countries as opposed to in a single one. Just such a situation might arise if for example subversive movements were to become active simultaneously throughout Western Europe at the same time as the Eastern Bloc threatened an invasion. In this case there are two fundamentally different ways in which the situation could be tackled. The first alternative would be for each country to accept responsibility for taking action against all subversive elements within its own boundaries, and the second would be for the alliance as a whole to take responsibility for handling subversion and insurgency throughout its entire geographical area, in the same way as it has responsibility for handling the external threat. In other words the alliance as a whole takes on the role of the host country.

Bearing in mind once more the essential nature of subversive war and its relation to a country's population, it would seem that the problem would have to be handled on a national basis, but the question then arises as to whether in fact it could be handled in this way from a strictly practical point of view. In the example quoted many men and a great deal of material would have to pass through the Belgian, Dutch and German ports and thence along Belgian, Dutch and German roads immediately preceding the outbreak of war if the British and American armies were to be ready to handle the attack, and if the Belgian and Dutch armies were to deploy properly. Under these circumstances it is questionable whether the NATO high command could afford to leave the matter solely to the national authorities, and it is even more questionable whether the national authorities could deal with the situation without diverting large numbers of men from NATO formations. Similar situations would be cropping up all over the affected area and the danger would not only concern ports, roads and railways but would also include the working of industry, communications, power and water supply amongst other things. Furthermore, although an orthodox war could be expected to reach a climax within a matter of days, a subversive campaign combined with sabre rattling from across the Elbe could go on for weeks or months unless it was brought under control.

It is not intended to make concrete recommendations as to how this highly complex problem should be tackled by NATO because there is too little data available on which to base a judgement, and in any case the setting is only used as an example of the sort of situation which might occur. The point which has to be made is that problems of this sort could arise, either in Europe or elsewhere in the world, and it is necessary for any country which is preparing to go to war against insurgency to understand how military and civil activities can be tied together in situations of varying complexity some of which might even involve groups of nations in the role of the 'Host Country', groups of nations in the role of the providers of assistance, and the whole operation having to be undertaken in conjunction with a military deployment against conventional aggression or the threat of it. There is little doubt that correct decisions will only be made regarding the machinery which should be set up if the countries concerned have studied the problems in advance, and understand the principles. If decisions have to be made on the spur of the moment some vital factor is going to be overlooked and the enemy may well achieve his object without even using his conventional forces. There would seem to be a strong case for ensuring that individual countries and alliances such as NATO take account of the machinery needed for dealing with insurgents when conducting contingency planning relating to conventional war, and that bi-lateral staff talks should take place between nations which are likely to find their troops stationed in the country of an ally.

PART TWO

THE ARMY'S CONTRIBUTION

Chapter 4

The Preparatory Period

Hitherto the subject matter of this study has been concerned with the background to campaigns of subversion and insurgency. It is essential that military officers should understand the background, so that they can give the right advice to politicians and administrators, and so that they can themselves make the right decisions at any stage of the struggle. It is now possible to progress further and to examine the actual tasks which officers and men of the army are likely to be called on to undertake, either as individuals or as members of units and formations, and the purpose of this chapter is to examine the contribution which the Army should be ready to make during the period when the enemy is building up his organization before the start of the campaign proper.

Looking in retrospect at any counter-subversion or counter-insurgency campaign, it is easy to see that the first step should have been to prevent the enemy from gaining an ascendancy over the civil population, and in particular to disrupt his efforts at establishing his political organization. In practice this is difficult to achieve because for a long time the government may be unaware that a significant threat exists, and in any case in a so-called free country it is regarded as the opposite of freedom to restrict the spread of a political idea. This seems to apply even when the idea is communism which openly declares itself determined to stamp out freedom itself, and which may well finish up by ordering the country's foreign policy for the benefit of another power such as Russia or China, rather than in the interests of the people of the country concerned. In the very early stages of subversion the government is therefore unlikely to regard action against those involved as being a practical proposition, even if it is aware of the threat, and this attitude may continue for as long as the opposition restricts itself to spreading its ideas by peaceful means. In most cases there is unlikely to be any violence until the idea has caught on to some extent, because until there is a modicum of support in some areas, it would be too dangerous for anyone to indulge in the use of force. Only after the idea has been accepted by a proportion of the people can its further spread be accompanied by violence.

But although action against those involved in subversion may not be possible, preparations for dealing with the situation can, and should be made, and in this context the army should become involved as soon as a threat is detected, in an advisory capacity. In the Introduction to this book the point was made that it is the job of soldiers to know how to use civil as well as military methods for fighting subversion, because fighting is a military occupation and members of the civil administrations are not taught how to do it. This is as true at the national level as it is at the provincial or district level, and representatives of the armed forces should be brought into the business from the very beginning. There is no danger of political repercussions to this course of action, because consultation can be carried out in strictest secrecy. The danger of not doing it is that the situation will be looked at only by those personally concerned with governing or administering the country affected, and such people will not as a rule be qualified to interpret what they discover in the light of study and of experience gained over the years in different parts of the world. It is extremely difficult for people who have not studied subversion in detail to interpret the threat accurately during the all important early stages, because only a small proportion of the enemy's activities are likely to come to light at that time. If the information centres around one or two strong-arm groups it is all too easy for the civil authorities to imagine that they are dealing with a few bandits or gangs of teenaged thugs without realizing that they are looking at the tip of an ice-berg. The natural desire to avoid being an alarmist, together with the ingrained reluctance of administrators to spend money on such unproductive measures as preparations for countering disturbances which have not yet broken out, is enough to ensure that they will under estimate the threat.[1]

Perhaps the first bit of advice which the armed forces should give to the government is to set up machinery at the top into which further advice can be fed because unless a supreme council for dealing with the trouble is formed on the lines mentioned earlier, it is unlikely that sound advice will get to the right quarters or be acted upon resolutely. If command machinery can be set up at the lower levels so much the better, but for political reasons it may not be possible in the early stages. In any case it is a mistake for ma-

[1] JOHN MCCUEN, op. cit., p.35.

chinery to be set up at a lower level unless there is a supreme council above it, because if this happens, the members of the subordinate group will get instructions from their various superiors at the national level which may be contradictory and will at best be unco-ordinated.

Having got the machinery established, the next task for the military leadership is to present to the supreme council a number of issues of a joint military civilian nature on which firm policy rulings should be taken before operations against those practising subversion can start. An excellent example concerns the way in which the Law should work. Broadly speaking there are two possible alternatives, the first one being that the Law should be used as just another weapon in the government's arsenal, and in this case it becomes little more than a propaganda cover for the disposal of unwanted members of the public. For this to happen efficiently, the activities of the legal services have to be tied into the war effort in as discreet a way as possible which, in effect, means that the member of the government responsible for law, either sits on the supreme council or takes his orders from the head of the administration. The other alternative is that the Law should remain impartial and administer the laws of the country without any direction from the government. Naturally the government can introduce new legislation to deal with the subversion which can be very tough if necessary, and once this becomes law the legal services will administer justice based on it. But the resulting situation is very different from that described in the first alternative because in the second case the officers of the law will recognize no difference between the forces of the government, the enemy, or the uncommitted part of the population. Anyone violating the law will be treated in the same way, and the full legal procedure, complete with its safeguards for the individual, will operate on behalf of friend and foe alike. As a rule the second alternative is not only morally right but also expedient[1] because it is more compatible with the government's aim of maintaining the allegiance of the population. But operating in this way can result in delays which might be impossible to accept if, for example, it looked as if subversion was going to be used in conjunction with an orthodox invasion or with the threat of one. The system might also prove unworkable if it were found to be politically impossible

[1] ROBERT THOMPSON, *Defeating Communist Insurgency*, p.53.

to get sufficiently severe emergency regulations on to the statute book. The point of raising this matter here is not to discuss it for its own sake, which out of the context of a particular situation would be pointless, but to describe the sort of issue which ought to be put to the supreme political body by the military authorities after consultation with the other government departments concerned together with all the operational implications both civil and military before the start of operations. Unless this is done, policy on matters important to the outcome of the struggle will just grow up as opposed to being decided on consciously in the light of all the relevant factors.

A further example of an issue of this sort concerns the extent to which force should be used, either by the police or by the army. In this case the politicians will rightly want to avoid the use of force as far as possible, and for as long as possible, because of the adverse effect it is bound to have on public opinion in the world at large and at home. At the same time there are military difficulties about using too little force and about delaying its application for too long, and these difficulties which are fully discussed in the next chapter must be presented to the political leadership by the military authorities after consultation with the other parties concerned, particularly the police.

An interesting parallel to the advice which the politicians should be getting from the military during the preparatory stages of a counter-subversive campaign can be found in the advice which a potential Host Country might be getting from a United States Country Team during the period when a subversive organization was thought to be making ready to launch a campaign. The subjects discussed above, e.g. the setting up of machinery to co-ordinate action against the subversive group, the pros and cons of one legal system rather than another, the time of the first show of force and the extent to which it should be used, are exactly the sort of factors which the military member of the Country Team must sort out during the period in which the plan is being prepared under which aid will be passed to the Host Country. In itself this is not important but it emphasizes the point that one country at least, that it to say the United States, recognizes the significance of getting these matters discussed and resolved before troops are committed.

Moving away from the field of advice and co-ordination the next thing to consider is ways in which the army can take a direct part in helping to prepare for the struggle before military units are deployed, and in this connection one of the most useful methods of providing assistance consists of making individual officers and soldiers available to agencies which are obliged to expand rapidly. It has been shown that in the pre-violence period the enemy is likely to be occupied in spreading his cause by propaganda of one sort or another, and in clamping a tight control on the people in the form of a chain of committees and cells to ensure that they accept the message and are thereafter in a position to act against the government in furtherance of the cause. In order to counter these moves the government must first know about them in detail which means that it must build up and adapt its intelligence organization to meet the threat. It must also promote its own cause and undermine that of the enemy by disseminating its view of the situation, and this involves a carefully planned and co-ordinated campaign of what for want of a better word must regrettably be called psychological operations. Finally in some circumstances it may be necessary for the government to try and organize the population along lines similar to those employed by the enemy. Each one of the activities is largely dependent on the other two, and each one is primarily a function of the civil authority which of course includes the police. It is now necessary to look at each one in turn to see what sort of help and indeed impetus can be provided by the army.

In general, it seems to be extremely difficult to get countries to maintain effective domestic intelligence organizations in time of peace, not only because of the expense, but also because of the feeling that such establishments run counter to the concept of the freedom of the individual. There is of course an element of truth in the idea that an effective domestic intelligence system could be used to jeopardize the freedom of the individual if it fell into the wrong hands, but the danger posed by subversion unchecked by good intelligence is far greater. The right answer in a free country is to have an efficient intelligence organization in the hands of people who are responsible to, and supervised by, the elected government:[1] under an authoritarian regime freedom of the individual is not particularly relevant.

[1] JOHN MCCUEN, op. cit., p.114.

But even an efficient intelligence organization has got to expand and adapt itself to new circumstances, if a campaign of subversion starts. In normal times, and in the very early stages of subversion, the intelligence organization has got to be able to penetrate small and highly secure targets. This fact governs the way in which it is established and organized and it also governs the outlook of those employed in it. Quite rightly preference is given to the development of a small number of really good sources as opposed to a larger number of more superficial ones. While the target is small and political only a highly specialized and secure organization can hope to get useful results without itself being penetrated by the enemy. But as subversion develops, the target will rapidly get larger and less secure. Committees will proliferate and strong-arm groups will spring up which may later coagulate into larger bodies of armed men. In order to deal with these groups effectively, much more information will be needed. If it is not available the government will not be able to control the situation quickly and economically, which it must be able to do if the enemy is going to be defeated.

For the purposes of presentation it is worth making a distinction between the sort of information which the intelligence organization is required to produce in normal times, and that which it will have to get after subversion has started. To this end the first sort of information might be described as political intelligence, and the second sort as operational intelligence. Naturally the need for political intelligence does not cease merely because there is also a requirement for operational intelligence: it will continue to be required throughout the campaign and after it is finished, in the same way as it was before it started. Operational intelligence on the other hand, will automatically cease to be required once the enemy is fully defeated, because it is concerned with information about the enemy's forces and committees which will have ceased to exist by that time. The distinction between these two forms of intelligence is an important one and needs to be well understood because the methods of collecting one sort are so different from those involved in collecting the other sort. The problem of preparing an intelligence organization to deal with subversion and insurgency is not therefore merely one of expansion. Developing new methods to deal with the new requirement is just as important, and far more difficult.

The practical problems raised by the requirement to expand and at the same time to develop new methods of working, are enormous. While some members of the original organization continue operating against their old targets in the usual way, others have to develop the new ideas, but being themselves inexperienced and untrained in these new methods they will have great difficulty in training their recruits. At this stage a major difference of opinion is likely to arise between those who wish to develop methods of getting operational intelligence based on the old principle of a few good sources being better than a host of poor ones, and those who consider that a large number of low grade sources are essential from an operational point of view. The influx of extra members will also raise security problems which will be greatly aggravated if the new members start using large numbers of low grade sources. These problems are bad enough when only one intelligence organization exists in the country concerned, but in many countries, including Britain, there are several separate organizations in existence each of which may wish to have a finger in the pie.

The army becomes involved in the problem in two ways. In the first place it is directly concerned in the way in which the intelligence organization develops, because in the later stages of the campaign when its units are deployed, it will rely very greatly on the information provided by the intelligence organization for the success of its operations. In this connection it is not only interested in ensuring that information is made available to it, but it is closely concerned with the sort of information produced. For example a lot of low grade information is more use tactically than a small amount of high grade material for reasons which are fully explained in Chapter 6, and this means that from the army's point of view an intelligence organization which relies on a large number of low grade sources is more valuable than one which concentrates on a few high grade ones, although naturally enough military commanders are only too pleased if some high grade sources can be operated at the same time. The author found general agreement for this view amongst officers of several nations at a symposium on counter insurgency run by the Rand Corporation in Washington in 1962; Colonel Fertig of the US Army, experienced both as a guerilla and a counter guerilla being particularly strong in his support.[1] Roger Trinquier also stresses the need for

[1] RAND CORPORATION, *Counterinsurgency: A Symposium (R-412-ARPA)*, 1963, p.127.

large numbers of low grade sources.[1] Anyway, regardless of the rights and wrongs of this particular issue, it is up to the army to ensure that its views on how the intelligence organization develops are represented forcefully and in time.

The second way in which the army can sometimes become involved, is by providing individuals to reinforce the intelligence organization or organizations, and under certain circumstances it may even be necessary for the army to move into an area and set up an intelligence organization from scratch if none exists. In order to understand the problem which the army may have to face, it is necessary to look at the sort of intelligence organizations which may exist in the areas where the army might be deployed in a counter-subversive war.

As a rule, whenever Britain has been engaged in fighting subversion or insurgency, she has been careful to build up a single organization charged with the responsibility for getting all intelligence about the enemy, irrespective of whether it concerns armed groups, supporters cells or the political party which is at the root of the trouble. Furthermore information relevant to the campaign which is picked up by other intelligence or counter-intelligence organizations operating in the area, is fed into the system by having representatives of the organizations concerned sitting on a joint intelligence committee, chaired by an overall director of intelligence. The British system derives from the fact that in the past she has usually become involved in fighting insurgency in her own territories where Police Special Branches have existed. Under these circumstances a single organization system based on Special Branch is undoubtedly the best. But in the future, Britain may well become involved in foreign countries, and the question arises as to whether a single organization would necessarily be the best under these circumstances or even whether it would be possible.

The advantages of having a single intelligence organization are fairly obvious. With more than one agency, spheres are bound to overlap because the operational target is only an extension of the original political one, contradictory information is likely to be produced about the same target, sources start working for both organizations at the same time, and before long open warfare may even break out in the underworld between the followers of the

[1] ROGER TRINQUIER, op. cit., pp.35–37.

74

two factions. On the other hand there are advantages in setting up a separate organization to deal with the new commitment of providing operational information, two of which are worth mentioning. The first one is that whatever organization existed before the onset of the trouble can get on with its important job of collecting high grade political information without being disrupted by rapid expansion, without having to set about devising the new methods needed for getting operational information, and without having to worry about the threat to the security of its long term agents which would arise from the recruitment of a large number of imperfectly trained reinforcements. The second advantage is that if a separate organization is set up to collect operational information, people can be selected to man it who can develop the necessary technique in the light of the prevailing circumstances, uninfluenced by notions picked up over the years doing the different work involved in collecting political intelligence. Broadly speaking, if all other things are equal, the decision as to which system should be adopted will rest on what sort of intelligence organization exists at the start of the trouble, and what sort of people are available to reinforce it, or set up a new one. Needless to say if it is decided to employ two intelligence organizations, their output should be carefully co-ordinated, possibly by a committee presided over by a single director of intelligence.

But often the number of intelligence organizations will not be a matter for choice, especially if one country goes to the assistance of another one which itself operates more than one agency. In this case the sort of decisions required will relate to whether the ally should rely solely on the host country's intelligence service or whether the ally should set up its own separate service. Once again it is impossible to say for certain which is the better system because situations vary to such a great extent. Sir Robert Thompson makes the point[1] that any government which can not find enough trustworthy people to run a single intelligence service will not last long, but in practice rulers or ruling bodies often like to play safe by running several intelligence organizations. In some countries the army and not the police are responsible for the domestic security of the state and so have to have an intelligence service but at the same time the police have a separate one in order to keep a check on the army. In other places the army,

[1] ROBERT THOMPSON, *Defeating Communist Insurgency*, p.86.

working under the Minister for Defence, can be used as a balance against a powerful Minister of the Interior who has control of the police, in which case both are sure to have their own intelligence service.[1] If Britain found herself going to the assistance of such a country she would obviously have to adapt her intelligence contribution to the system in force, although she could and should use her influence to rationalize it as far as possible.

At this point it might be of interest to mention a few instances of multiple intelligence organizations in order to demonstrate how frequently they are found in practice. In the Philippines the following separate intelligence agencies all functioned in Manila at the same time and were held to have produced good results at a low cost. The Intelligence Service of the Armed Forces of the Philippines, the Counter-Intelligence Service of the Armed Forces of the Philippines, the Intelligence Service of the Philippines Constabulary, the National Bureau of Investigation, the Special Branch of the Manila Police, and private intelligence organizations belonging to the President of the Republic, Internal Revenues, Customs and Immigration. From about 1950 a Military Intelligence Service co-ordinated all operational intelligence activities but the other agencies continued to operate for their own purposes.[2] In Algeria the French army was responsible for all intelligence work outside the cities using its own Military Intelligence Service. Inside the cities the job was done by three separate agencies, the Police de Renseignements Generaux, the Bureau de Securité de Territoire, and the Suréte.[3] In Saigon in 1966 there were apparently no less than seventeen intelligence agencies at work,[4] and even in Aden ten separate organizations were operating until Brigadier Cowper was sent out in 1965 to re-organize the system.[5] By contrast in Muscat and Oman in 1958 no intelligence organization existed at all and one had to be hurriedly built up to support the offensive launched by the Sultan's forces, in conjunction with the British, against the rebels on the Jebel Akhdar.

The only general point which can be made about intelligence is

[1] RAND CORPORATION, op. cit., p.107.
[2] Ibid., p.108.
[3] Ibid., p.107.
[4] ROBERT THOMPSON, *No Exit from Vietnam*, p.124.
[5] J. PAGET, *Last Post : Aden 1964–67*, p.129

that it is of the very greatest importance, and regardless of the difficulties, any country involved in fighting subversion or insurgency should concentrate on setting up, or building up, the necessary agency or agencies. This is just as important if the operation is being carried out by an ally as it is if it is being carried out by the original country. The army has an important part to play in the business, firstly by ensuring that intelligence is directed in such a way that its own requirements are met, and secondly by making a direct contribution in terms of men and material as required, even if, as sometimes happens, the body operating the existing organization, e.g. the police, does not welcome the assistance. The army is also responsible for providing liaison officers in order to ensure that information gathered by the intelligence organization reaches it, but this is a very unimportant matter by comparison with its other responsibilities. It is useless to imagine that the provision of liaison officers is all that is required.

The next area in which the army can make a contribution before the outbreak of violence lies in the field of psychological operations and propaganda, where the government not only has to counter the steps which the enemy are taking to get their cause across to the population, but also has to put across its own programme in an attractive way. There are three aspects to this business. In the first place careful assessments and appreciations have to be made by trained men and presented to the government leadership at the various levels so that policy can be laid down. This policy then has to be turned into specific propaganda material such as films, broadcast programmes, newspaper articles, leaflets and so on. Finally the material has to be disseminated by mechanical means, that is to say by broadcasting, printing, or by the projection of films on the screens. In the defensive context mechanical devices are also required for locating illegal enemy broadcasting stations and for jamming them, and for monitoring enemy propaganda so that it can be correctly countered.

In order that these three functions can be carried out, a psychological operations organization is required analogous in a sense to the intelligence organization, although it need not be nearly so large. This organization should be planned on the basis that it must provide operations teams at every command level responsible for drawing up assessments for the benefit of the appropriate committee or commander and responsible also for translating that policy into specific material. The head of each team would of course be the psychological operations adviser to the committee or commander concerned. In addition to these teams a number of psychological operations units of various sorts will be required whose job is to handle the mechanical processes involved in detection and dissemination, and which can be sent to local areas to work under the direction of the appropriate team if required, but which would more normally work under central direction.

The main difference between the psychological operations organization and the intelligence organization in the early stages, is that whereas there is likely to be at least one intelligence organization in existence before the onset of the campaign on to which reinforcements can be grafted, there is unlikely to be any corresponding organization dealing with psychological operations. It is all the more important therefore that the government should be in a position to set one up at short notice, because so much depends upon it. All too often successful government action in the civil and military field is rendered completely useless because the machinery for exploiting success in the minds of the people is non-existent. At the same time the enemy who have suffered a reverse in fact, are able to nullify it, or even turn it to their advantage in the minds of the people, because they have the means of getting their version of events across. It is only necessary to stress once again that wars of subversion and counter subversion are fought, in the last resort, in the minds of the people, for the importance of a good psychological operations organization to become apparent.

It is not possible in the context of this study to discuss in detail the ways in which the psychological operations organization should work, any more than it is possible to discuss how the intelligence organization operates, but two points are worth emphasizing. The first one is that although the government may provide the framework and the leadership for the psychological operations organization from outside the country in which the campaign is being fought, e.g. from Britain in Aden, it must incorporate individuals from the country itself to advise on the form which the propaganda should take. No matter how well a person from outside the country may think that he knows the way in which the minds of the local people work, he is none the less likely to make mistakes when preparing propaganda for them. This fact was demonstrated forcefully in Malaya as a result of a process of trial and error which lasted over several years.[1] The second point is that there is no particular reason why the army should be responsible for setting up the psychological operations organization, so long as some body is organized in advance to do it at short notice: there must be no delay once the threat has been recognized. In fact in Malaya the psychological operations organization and the information service, i.e. that part of the organization which directed its activities at the uncommitted civilians as opposed to the enemy, was essentially an organ of the civilian administration, but it grew up over a period of years after the emergency started, and was not able to operate effectively in the very early stages, when it was so sorely needed. But in practice the army is the only body which can maintain in peacetime a nucleus capable of going to a country at short notice so as to build up a complete service, and it is probably true to say that if the army does not take on this commitment, no one else will. At the very least the army is likely to have to produce individuals for the organization and some of the specialized units.

The last of the three main ways in which the army can contribute to a counter-subversive campaign in the early stages concerns the process of counter organization, which is a term used to describe a method by which the government can build up its control of the population and frustrate the enemy's efforts at doing so. In its simplest form, counter organization involves putting the government's views over to the population by action rather than by propaganda. For this purpose individuals can be sent amongst the community for the purpose of doing work which will help to remove sources of grievance and at the same time making contact with the people. The sort of jobs which can be undertaken range from teaching to the setting up of clinics, advising on simple construction works and working on agricultural projects. Undoubtedly it is desirable that civilian administrators should be used if they are available, but as a rule they can not be found in sufficient numbers. In this case officers and men of the army must be used instead, and even specialist units can be provided on occasions which enables more ambitious projects to be undertaken such as the building of roads and bridges. The concept of military aid to the civil community is not new to the British Army, an excellent

[1] RICHARD CLUTTERBUCK, op. cit., p.106.

example of it being the use of Royal Engineer detachments in Anguilla. The French also used the system in Algeria, and so did the government of the Philippines during the rebellion there.[1] In the United States Army the carrying out of this task is the responsibility of commanders at every level who are supposed to use whatever resources they can spare to assist the population in helping themselves. Sometimes 12-man Special Forces Teams or Marine squads in the Combined Action Programme have been used in village communities on civic action projects.

In Algeria the French developed the system to a far greater extent by sending out teams into the towns and countryside whose job was to set up a complete chain of committees and cells supporting the government on similar lines to those established by the enemy.[2] In this way they got right under the skin of the population, and by introducing identity cards, branding livestock with the identity card numbers of their owners, and by other similar methods they soon imposed a tight control over the people. As the system developed the influence of the government increased. The committees provided a framework which helped to engender a feeling of security and commitment amongst the people which in turn encouraged them to give information. This enhanced still further the power of the teams which were able to impose effective sanctions or work for the benefit of the community, according to the extent to which the people were supporting the government. In short the system greatly increased the power of the government in the areas where it was working.

It is probably true to say that if teams could be got into operation during the early phases of subversion, the enemy would never have a chance of building up his organization at all, but there is no record of the system being introduced sufficiently early for this to have happened. Certainly it would be unthinkable for the British to operate teams in this way before any violence had taken place because they could not function without legal backing, and it would be politically difficult if not impossible to pass the necessary legislation until something had happened. For this reason it is unlikely that the British army could be persuaded to allocate much in the way of resources towards maintaining teams of this sort in advance of a requirement arising although, as already

---

[1] JOHN MCCUEN, op. cit., pp.96–97.
[2] ROGER TRINQUIER, op. cit., p.30.

mentioned, the United States Army does so. None the less the British Army could well afford to study the problems involved, and have suitable men earmarked to take part in an operation of this sort. If this were to happen teams could be formed from earmarked people as soon as the threat received recognition, and they would then be ready to operate by the time it became politically possible to take the necessary action and deploy them. This situation is of course totally different from that governing the deployment of intelligence resources and psychological operations teams who should start to operate as soon as the threat is recognized.

In summing up the contribution which the army can make before the onset of violence it is only necessary to point out two main ways in which it differs from other government agencies. The first of these is that it has a fund of knowledge of methods which can be employed to counter subversion and insurgency gained by study and experience. The second is that it can train and maintain individuals and units in advance of a situation arising who can move at short notice to an affected area and start working at once. The extent to which it is effective in these respects is of course dependent on the amount of effort and resources which it is prepared to expend in preparing for the task. There is no doubt at all that the outcome of a campaign of this sort depends very largely on the action taken in the early stages, and for this reason the army's ability to use its unique characteristics in this period is of immense importance. It is perhaps unnecessary to point out also that the foundations laid at this time will either promote or bedevil all that follows. It is for example much more difficult to turn an ineffective intelligence organization into an effective one at a later stage in the campaign than it is to build up a good one in the first place.

Chapter 5
The Non-Violent Phase

Gene Sharp claims that 125 different forms of non-violent action have been identified which he classifies under three heads as non-violent protest, e.g. pickets and street corner meetings; non-co-operation, e.g. strikes and boycotts; and non-violent intervention, e.g. sit-ins and various forms of obstruction.[1] He also lists 84 examples from history starting in 494 B.C. showing briefly which techniques were employed on each occasion and the purposes for which they were used.[2] In considering the nature of campaigns of non-violence two points are at once apparent. The first one is that the initiators of non-violent action have not always been opposed to the use of force on moral grounds, but have often employed non-violence because they considered it to be better suited to the prevailing circumstances,[3] with particular reference to the credit which they thought might accrue to them in terms of public opinion. On occasions non-violent action has been used in conjunction with a campaign of terrorism, sabotage or full-scale insurgency, and even when the organizers have themselves been opposed to violence on moral grounds they may, like Gandhi, have been prepared to make common cause with those who have no such inhibitions.[4] The second point which is apparent from the study of past campaigns is that non-violent action has on occasions become violent as a result of a march or meeting getting out of control either because of an accident or because large bodies of people mobilized for non-violent action fell under the influence of those whose aim was to use them in a different way.

In view of the prevalence of non-violent action in the world today, it is necessary to point out that a great deal of it does not warrant the description of subversion at all, because it is not seriously designed either to overthrow the government, or to force concessions out of it. Much of it takes place in the context

---

[1] GENE SHARP in *The Strategy of Civilian Defence*, edited by Adam Roberts, Faber and Faber, 1967, p.89.
[2] Ibid, pp.100–103.
[3] GEORGE WOODCOCK, *Civil Disobedience*, CBC Publications, 1966, p.5.
[4] Ibid., p.44.

of demonstrations of solidarity for various causes, and some of it, especially that which concerns the activities of university students, is best described as educational in so far as it is designed to fit those practising it to take part in revolutionary work later on in life,[1] or even as recreational in so far as the participants, main concern is with the enjoyment which they get out of it.[2] Another reason suggested for the popularity of student revolt is that those taking part are activated by a desire for the 'comradeship and manliness' engendered by participation in revolutionary activities,[3] but whether or not this is so, the results are not as a rule sufficiently serious to warrant the description of subversion.

In practice a campaign of genuinely subversive non-violence usually occurs in one of three contexts. Either it is intended to achieve the aims of the organizers on its own without ever becoming anything but non-violent, or it is intended to be used in conjunction with full-scale insurgency elsewhere in the country in which case its purpose is likely to be to divert soldiers from the main battle, to involve the government in extra expense and to attract the attention of sympathetic world opinion, or it is part of a progression towards a fully developed campaign of urban terrorism and sabotage. In this connection the point has already been made[4] that civil disorder is one of the recognized phases in the development of insurgency in an urban area, and non-violent activities are essentially geared to urban life because they rely on the participation of large numbers of individuals.

The purpose of this chapter is to examine the contribution which the army must be prepared to make as part of the overall government effort to overcome non-violent subversion. In common with its role in the organizational phase mentioned in the preceding chapter, its contribution can conveniently be considered first in relation to its responsibilities for advising the civil leadership of the country, and secondly in relation to the operations which it can itself carry out either in conjunction with the police or separately.

[1] HOCH and SCHOENBACH, *The Natives are Restless*, Sheed and Ward, 1969, p.203.
[2] DAVID DONNISON in *Students in Conflict*, Weidenfeld and Nicolson, 1970, p.XXIX.
[3] MARTIN OPPENHEIMER, *Urban Guerillas*, Penguin, 1970, pp.54–55.
[4] See p.38 above.

Perhaps the first thing which an army officer responsible for advising the civil authorities should do, is to try and recognize the weak points in the enemy's position so that they can be exploited in the overall government plan. Naturally these weaknesses will depend mainly on the circumstances of the case, but there are three inherent ones which almost always appear in connection with programmes of non-violent action. The first of these relates to the problems which face the organizers in getting their campaign started, the second to the difficulties which face them in keeping it going, and the third to the machinery which has to be set up in order that it may be controlled. It is worth examining each of these weaknesses in turn.

Terrorism, sabotage, and even guerilla action, only require the active participation and support of a relatively small number of people in the early stages, and this can often be arranged by indoctrination and threats in the preparatory period. Non-violent action on the other hand, involves persuading a large number of people to do something positive such as take part in a mass meeting or a protest march or go on strike. But it is rare to find large numbers of people who are so interested in a political cause that they are prepared to abandon their work and sacrifice their recreational time merely to stand around in a group being troublesome to the government on the off chance that it will make concessions in some direction which will probably bring them little personal benefit or satisfaction. In fact only the hard core organizers are likely to be sufficiently dedicated to behave in this way, and such people are normally viewed with suspicion by the average working man or housewife and even by the majority of the student population. Experience on both sides of the Atlantic shows that the hard core is only likely to be successful in mobilizing large numbers of people if it can make use of an intermediary section of the population who might best be described as politically conscious idealists. There is usually a small number of such people in any community and although they are likely to be far more moderate than the organizers of the subversion, and although they are most unlikely to want to overthrow the government, they can often be persuaded to take part in a demonstration of some sort for a limited objective. If they can once be got onto the streets, even in relatively small numbers, it may be possible for the extremists to goad the authorities into taking some violent action against the moderates which will at least attract the sympathy of the uncommitted part of the population, some of whom may even align themselves with them. As numbers increase, clashes with the authorities will increase also, and more and more of the population can be expected to become supporters of the campaign. The procedure is very cumbersome but if the government fails to understand what is going on and falls into the trap, it may well succeed. On the other hand if the government realizes the position, it should be able to win over the moderates and isolate the extremists by exposing their motives and by discrediting them.

In one sense the problems involved in keeping the campaign going are no more than an extension of those concerned with getting it started. Even if the government fails to isolate the extremists from the population in the first instance, it should be able to drive a wedge between the two groups later on, because of the divergence which is likely to exist between the aims of those organizing the campaign and the cause or causes which they have to put forward in order to attract support. The significance of causes in relation to subversion and insurgency as a whole, was discussed in Chapter 2, where it was shown that although the organizers can, and often do have to camouflage their true aims in order to make them acceptable to a sufficiently large sector of the community, they can not afford to present a cause which actually runs counter to them. But in order to promote a campaign of non-violence this is often what has to happen. As explained above, the aims of the organizers may be to overthrow the government but actual marches, strikes and rallies can only be arranged on the basis of an appeal to the government to grant some particular concession. Only in the case of protest against foreign occupation is it likely that sufficiently large numbers could be got together other than in the context of a specific appeal. As the campaign develops, a split is likely to open between the organizers and their followers, and the more successful the campaign the wider will be the split, because the greater the number of concessions granted by the government, the less have the participants to gain from seeing it overthrown.[1] Furthermore if the campaign achieves no success, the organizers will have to adopt more forceful methods and the non-violent nature of the operations will gradually

[1] MARTIN OPPENHEIMER, op. cit., pp.117–118.

diminish. The disturbances which took place in France in May 1968 afford a good example of a campaign which was successful from the point of view of the people, but unsuccessful from the point of view of the organizers. In this case the leadership wanted to trigger off disturbances which would overthrow the French government and substitute for it a non-bureaucratic form of revolutionary communism. But this could hardly be expected to appeal to the workers, or to the students who had been selected to act as the vanguard of the revolt. As a result the cause was presented to the workers as a demand for more pay and to the students as reform of their university. Both of these causes were good ones but they ran counter to the aims of the organizers because as soon as the necessary concessions were made the revolt petered out. Neither group wanted a revolution which would have robbed them of the fruits of their labour and of their study respectively. Vittorio Rieser, an Italian student leader, expresses the dilemma well by pointing out that the organizers of student movements and other minority groups are trapped in a sort of vicious circle in which ultimately the only object of action is action itself.[1] In short, except where the real object of the campaign is related to the expulsion of an occupying power, it carries the seeds of its own destruction within it.

There are two main difficulties which confront the organizers of a non-violent campaign when it comes to controlling their followers, both of which are capable of being exploited by the government. The first of these is that a large number of people have to be involved compared, for example, to the numbers required to conduct a programme of sabotage or terrorism, and the second is that the participants themselves are not disciplined members of a clandestine organization, but crowds of citizens or groups of students who may resent tight, political organization as part of their beliefs.[2] Even when the participants are united in their opposition to the governing authority, control presents a major problem. In his book about the Nazi occupation of Denmark, the SOE[3] leader Flemming Muus describes how a Freedom Council

---

[1] VITTORIO RIESER in *New Revolutionaries, Left Opposition*, edited by Tariq Ali, Peter Owen, 1969, p.205.
[2] COHN-BENDIT, *Obsolete Communism, The Left Wing Alternative*, Andre Deutsch, 1968, p.250.
[3] SOE, Special Operations Europe.

was set up in June 1943 to co-ordinate violent and non-violent resistance[1] and how this was subsequently streamlined in May 1944,[2] but in conversation with the author of this study he stressed the limitations imposed on the non-violent side of the programme by the difficulty of controlling it, and by its vulnerability to enemy penetration and ruthless counter-measures.

With the major weaknesses of non-violent action in mind, it is possible to consider a general framework of operations suitable for combating it. For the purposes of this study no account will be taken of the simplest method of all, which is to suppress the movement by the ruthless application of naked force, because although non-violent campaigns are particularly vulnerable to this sort of action, it is most unlikely that the British government, or indeed any Western government, would be politically able to operate on these lines even if it wanted to do so. In practical terms the most promising line of approach lies in separating the mass of those engaged in the campaign from the leadership by the judicious promise of concessions, at the same time imposing a period of calm by the use of government forces backed up by statements to the effect that most of the concessions can only be implemented once the life of the country returns to normal. Although with an eye to world opinion and to the need to retain the allegiance of the people, no more force than is necessary for containing the situation should be used, conditions can be made reasonably uncomfortable for the population as a whole, in order to provide an incentive for a return to normal life and to act as a deterrent towards a resumption of the campaign. Having once succeeded in providing a breathing space by these means, it is most important to do three further things quickly. The first is to implement the promised concessions so as to avoid allegations of bad faith which may enable the subversive leadership to regain control over certain sections of the people. The second is to discover and neutralize the genuine subversive element. The third is to associate as many prominent members of the population, especially those who have been engaged in non-violent action, with the government. This last technique is known in America as co-optation[3] and is described by Messrs Hoch and Schoenbach as drowning the

---

[1] FLEMMING MUUS, *The Spark and the Flame*, Museum Press, 1956, p.107.
[2] Ibid., p.150.
[3] MARTIN OPPENHEIMER, op. cit., p.152.

revolution in baby's milk.[1] In a slightly different sense the idea of placing a personal obligation for helping to keep the peace on every adult male in the community has been a feature of English methods since Saxon times.[2]

But although action along these lines is easy to describe it may be extremely difficult to carry out. For one thing it is not always easy to make concessions to the extent necessary for separating the subversive leaders from their followers, because to do so in a democratic country usually means taking something from the majority to give to the minority: if it were the other way round the majority would have helped themselves already by normal Parliamentary means. Martin Oppenheimer expresses the dilemma using the term 'left' to denote the minority and 'right – or the CIA' to denote the majority.[3] Northern Ireland affords another good example at the time of writing, because certain sections of the Unionist majority are trying to prevent their own government from giving concessions to the Nationalist minority. This process of give and take is essentially a political matter but it is undoubtedly the duty of military leaders to stress the importance of reaching agreement quickly, especially if a pause has been achieved by the use of troops because, if the breathing space is lost, it may not be possible to restore the situation without a much more lavish use of force. Over the matter of identifying and neutralizing the genuine subversive elements, the army may be of more direct assistance because it can help to build up the intelligence organization as described in the previous chapter. Even if direct help is not required, it is the responsibility of the military leadership to ensure that the government affords the necessary priority to the task.

One last point concerning the overall conduct of the campaign is worth making which is that it may be disadvantageous to destroy a non-violent subversive movement if the only result is to find that a campaign of terrorism or all-out insurrection has been substituted for it. It is far easier to penetrate a subversive movement when it is using non-violent means than it is when the movement is wholly clandestine because of the number of people overtly involved. Although this can not be accepted as a reason

[1] HOCH and SCHOENBACH, op. cit., p.182.
[2] T. A. CRITCHLEY, *The Conquest of Violence*, Constable, 1970, p.28.
[3] MARTIN OPPENHEIMER, op. cit., p.91.

for containing it in this form indefinitely, it is worth considering whether it should be broken up before some headway has been made towards discovering the identity of the people who are really behind it.

Having decided on the general lines along which the campaign should be run, the problems become more directly operational, and one of the first things that has to be considered is the question of when, and in what form, force should be applied. Clearly there will be no question of using troops while the regular police are capable of controlling the situation by themselves, but the handling of mass meetings and protest marches, especially if they develop into riots, is very expensive in manpower, and the moment may well arrive when the police can no longer manage on their own. When this happens there may be a choice between using regular troops or using some sort of auxiliary organization such as army or police reservists, and army officers must be prepared to advise on this problem. Some of the factors which should be considered are given below. There are considerable disadvantages involved in employing reservists. For example they are probably less well trained and disciplined than regular troops, they are unlikely to have sophisticated equipment especially in the communication field, and they are not as a rule organized to operate far from their homes. Above all they are almost certain to be drawn directly from the population which has to be controlled, and may therefore be partisan in relation to the issues at stake and they are in any case subject to pressure applied on them through their families or through their civilian jobs. Although regular troops do not suffer from these disadvantages, they are much more expensive to use, especially if the emergency is protracted to the stage where it is necessary to provide reliefs for the units initially committed. Furthermore soldiers may be in short supply, especially if the non-violent campaign is part of more widespread disturbances, or if it takes the form of a protest at the employment of forces in some other part of the world. Finally the government may be reluctant for political reasons to use troops while there is a chance of the disturbances being controlled by other means. Whilst acknowledging the force of all these considerations it is none the less true to say that unnecessary delay in committing the army may result in far worse complications later on.

If it is decided that soldiers should be employed, the next

question concerns the amount and type of physical force which they should be authorized to use and in this connection the key factor is that the number of troops required to control a given situation goes up as the amount of force which it is politically acceptable for them to use goes down. For example three or four times as many troops might be required if they were only allowed to use persuasion, as would be needed if they were allowed to use batons and gas; and three or four times as many troops might be needed if they were restricted to using batons and gas, as would be required if they were allowed to use small arms. Undoubtedly the decision in this matter is a political one, and soldiers must be prepared to do their best in the light of whatever decision is made. It is none the less a matter of military judgement to work out how many soldiers will be needed to control a particular situation using a given level of force, and it is up to military commanders and staffs to ensure that government leaders know and accept the implications in terms of providing the requisite number of troops to carry out the task when they make the decision on the amount of force which may be used. In this connection it is also important that military commanders should point out the morale effects on the soldiers of operating in accordance with the level of force decision, because this is relevant to the length of time men can be used without relief and also to the extent to which they will require retraining after being relieved before being used in a genuine operational role. This factor is often ignored because military discipline enables risks to be taken in the short term, but it is a matter which has worried many regimental officers in recent years as a result of their experiences in such places as Cyprus and Northern Ireland. It would appear that something of this sort was behind the differences which Colin Mitchell had with his superiors in Aden[1] and it is certainly a problem which merits more attention than it normally receives. Finally, however great the restrictions imposed on the use of force by soldiers, every effort should be made to retain the respect and awe of the civilian community for the ultimate in terms of force which they might use. If an impression can be built up that although the troops have used little force so far, they might at any moment use a great deal more, the people will be wary and relatively fewer men will be needed. Admittedly

---

[1] COLIN MITCHELL, *Having Been a Soldier.*

an element of bluff is involved and one of the most difficult tasks facing a military commander is to get the maximum value out of it without having the bluff called. But risks have to be taken when countering non-violent subversion in the same way as they do in other forms of warfare, and it is up to the officers of the army to study carefully how best to take them.

A factor which is related to the attitude of the population towards the troops, concerns the extent to which the soldiers should be allowed to, and indeed encouraged to mix with the people, especially when off duty. This is an important matter because not only is it relevant to the respect in which the people hold the troops but it is also closely connected with the extent to which it is safe to expose the troops to possible indoctrination by subversive elements. Wolfgang Sternstein,[1] Sir Basil Liddell Hart,[2] and Adam Roberts[3] all touch on the danger involved, and although it need not be overstated when well disciplined troops are operating in a good cause, it is certainly a subject which commanders at all levels have got to consider. The risk is of course much less serious if the non-violent campaign is combined with measures involving loss of life amongst the men themselves. As usual no ruling can be given on the subject outside the context of a specific situation, but it would probably be true to say that the balance of advantage both in terms of retaining the respect of the population and in preventing subversion of the soldiers would lie in restricting the contact which the one should have with the other were it not for one further factor. This factor concerns the importance of gathering all possible information about the situation and the necessity for getting every individual soldier to play a part in collecting it.

Good information is an essential requisite for making the best use of a limited number of soldiers or policemen in countering non-violent subversion, as indeed it is for dealing with all forms of insurgency. In the early stages of a non-violent campaign, when demonstrations and riots are rife, information is much easier to obtain than it is once conditions become more settled. At this time the real leaders are likely to be around, possibly even negotiating with government forces as they become committed to action: at

[1] *The Strategy of Civilian Defence*, p.131.
[2] Ibid., pp.208–209.
[3] Ibid., p.243.

this time too members of the population who are frightened by the unexpected tumult may well talk freely in return for nothing more substantial than a bit of reassurance. But once the situation becomes more stable leaders will disappear underground and be replaced by 'front' men, and the ordinary people, freed from the stresses of imminent peril or possibly inured to them, will become more reserved. It is not unusual to find that the intelligence organization is unable to exploit the favourable situation which exists in the early stages, either because none exists or because it is too busy expanding and adapting itself to the new circumstances. It is therefore essential that soldiers and policemen should be trained to get all the information they can by overt means and their employment, and leisure activities, if any, should be planned with this in mind. Such a programme will only be effective if the officers understand through study the problems involved and are prepared to give effective direction. Arrangements must also be made for recording and passing on everything that is received. One of the main factors involved is continuity, and a platoon or company which can stay in the same region for a long time is worth several times as many men who are constantly moved from one place to another, because of the contacts and background knowledge which the stationary troops can build up in a particular area.

The factors relevant to the command and control of operations directed against non-violent subversion are essentially the same as those considered in Chapter 3. One point is, however, worth mentioning and that is that the government may not wish to acknowledge the seriousness of the situation in the early stages to the extent of setting up emergency committees which cut across normal local government authority let alone institute a single commander system for runnning operations. This might make for difficulties in co-ordinating the activities of the military and police, especially in a country like the United Kingdom where the official local government leader would be the part time elected Chairman of the Council concerned who might be totally unable, unsuitable or unwilling to act as the chairman of an operational committee. Another difficulty would result from the fact that in many areas police and local government boundaries do not coincide. For these reasons military commanders must be prepared to make *ad hoc* arrangements which might include the setting up of committees consisting of military and police officers only, such as the County Security Committees established in Northern Ireland during the latter part of 1969. Alternatively it might be possible to work in conjunction with the clerks of the various councils who could co-ordinate the activities of all concerned in the same sort of way as they sometimes do in the face of natural disasters. Any commander who really understands the principles involved in co-ordinating civil and military activities against sub-version should be able to devise a system providing that he also appreciates the necessity for having one.

All the points brought out in this chapter so far must be under-stood by army officers if they are to carry out their functions of advising the civil authorities and play their part in directing operations. These points must therefore be taught at the appropri-ate military training establishments and colleges, and the teaching of them represents a most important aspect of preparing the army to take part in this sort of operation. By comparison the business of teaching the soldiers actual military techniques such as methods of channelling protest marches or of handling riots is easy. These techniques are well known, the two basic functions being those of observation over a wide area followed by concen-tration and action if necessary at critical places. There are a few technical problems such as those relating to communication with the police but these can be solved quickly enough providing that the government is prepared to spend the money. The only other point that needs making, is that the government should be in a position to keep the country running during a prolonged period of strikes and civil disturbance, and to this end either the police or the army should have men available who are capable of operating essential services such as power stations and sewage systems with relatively little assistance from civilian experts.

In conclusion it is probably fair to say that a subversive cam-paign run on non-violent lines is relatively easy to handle. In theory all sorts of awkward situations can be envisaged which appear to present the government with no option other than that of surrendering or of resorting to a politically or morally unac-ceptable level of brutality. But these situations usually fail to take into account the built-in weaknesses of non-violence. Certainly the government must understand how to combat non-violence, and in particular how to exploit the weaknesses inherent in this type of conflict. It is up to the officers of the army to provide the necessary advice and expertise in this matter. Naturally too, the police and the army must be well trained in the techniques involved and they must be adequately equipped. Perhaps the most important single factor concerns the methods adopted for gathering good information and the army has an important part to play in this process as described earlier in the chapter. But if the government is efficient in these matters it seems unlikely that non-violent action will achieve much except in the case of the occupation of a country by a foreign state which is not prepared to use sufficiently ruthless counter-measures for one reason or another. Non-violent action in conjunction with sabotage, terrorism or full-scale guerilla war may well be effective as a means of presenting the cause to world opinion and it may be a useful method of diverting the efforts of the government's forces and of incurring them in extra commitments or expense, but in this case it constitutes no more than a particular tactic in a campaign which will be decided on its overall merits.

Chapter 6

INSURGENCY Part I:
Tactics: The Handling of Information

In the final phase of a revolutionary campaign as envisaged by Mao Tse Tung, armed insurgents come out into the open and fight the forces of the government by conventional methods, but in the earlier stages the war is fought by people who strike at a time and place of their own choosing and then disappear. Sometimes their disappearance is achieved by the physical process of movement into an area of thick cover such as a jungle, and at other times by merging into the population. In either case those who are supporting them by the provision of money, food, recruits, intelligence, and supplies rely for their security on remaining anonymous. The problem of destroying enemy armed groups and their supporters therefore consists very largely of finding them. Once found they can no longer strike on their own terms but are obliged to dance to the tune of the government's forces. It then becomes a comparatively simple matter to dispose of them. The purpose of this chapter is to examine the part which the army may be called upon to play in the process of finding and destroying enemy groups in order to point to the steps which should be taken in advance to prepare it for the task.

If it is accepted that the problem of defeating the enemy consists very largely of finding him, it is easy to recognize the paramount importance of good information. As a rule those taking part in counter-insurgency operations do quickly recognize it, and stress the need for a good intelligence service to be built up. It is even common to find commanders at every level laying the blame for such failure as may have attended their efforts on the shortage of good information given to them. But there is a fatal flaw in the thinking of those who put down their lack of success to the shortage of information given to them, which is that the sort of information required cannot, except on rare occasions, be provided on a plate by anyone, not even by the intelligence organization. If there was a system whereby an intelligence organization could do this, it would have been devised years ago, and there would be no such thing as insurgency because enemy armed groups and their supporters would at once be found, harried, tracked down and destroyed by the army and the police.

The fact is that although intelligence is of great importance, it does not usually come in the form of information which will immediately enable a policeman or soldier to put his men into contact with the enemy. The reason for this is inherent in the way in which intelligence organizations work, collecting information as they do by operating informers and agents, or by interrogating prisoners to mention only a few of their methods. Information collected in this way is immensely valuable for providing data on which policy can be worked out, and it forms the background to operational planning. But only occasionally can it be used to put troops directly into contact with the enemy because material about enemy locations and intentions is usually out of date before it can be acted upon by the soldiers. The sort of information which intelligence organizations produce has to be developed by a different set of processes to those used by the organizations themselves before it can be used for putting government forces into contact with insurgents. A cow can turn grass into milk but a further process is required in order to turn the milk into butter.

Two separate functions are therefore involved in putting troops into contact with insurgents. The first one consists of collecting background information, and the second involves developing it into contact information. To over-simplify the full process it could be said that it is the responsibility of the intelligence organization to produce background information and that it is then up to operational commanders to develop it to the extent necessary for their men to make contact with the enemy, using their own resources. Undoubtedly this is an over-simplification because, as will be shown, operational units have a function in producing background information, and the intelligence organization can help in the business of developing it, but it is absolutely necessary to understand the fact that the main responsibility for developing background information rests with operational commanders and not with the intelligence organization. Once this fact is accepted it is possible to look at the two functions involved and view the straightforward military techniques which have been developed over the years in their correct perspective. No matter how proficient soldiers and policemen become at using these techniques, they will achieve no more than isolated successes unless they can use them as part of an effective system for handling information.

All actions designed to retain and regain the allegiance of the population are relevant to the process of collecting background information because its provision is closely geared to the attitude of the people. In other words the whole national programme of civil military action has a bearing on the problem. The production of the national plan has already been discussed, and is in any case of too wide a scope to be examined further in this chapter. The operational responsibilities of a commander or committee at a low level, e.g. Province or District, for the collection of background information are usually concerned with affording a measure of security to the loyal and uncommitted sections of the population on the one hand, and organizing it on the lines discussed in Chapter 4 on the other. The army's contribution can best be summarized under three headings. First there is the advice which the military commander in each area should give regarding the extent to which government forces should be used on protective tasks as opposed to offensive ones, and there is the actual contribution which his men make in this respect. Second, there is the contribution which the army makes towards organizing the population along the lines suggested if this task can not be undertaken by the police or by some other government body. Third there is the actual background information which the army itself supplies from time to time in the form of documents or prisoners for example. A complete illustration in the form of a scenario is given in the later part of this chapter which goes through the processes involved both in collecting background information and in developing it into contact information.

It is now necessary to try and indicate how background information can be developed by operational units into contact information. Basically the system involves a commander in collecting all the background information he can get from a variety of sources including the intelligence organization, and analysing it very carefully in order to narrow down the possible whereabouts of the enemy, the purpose being to make deductions which will enable him to employ his men with some hope of success as opposed to using them at random in the hope of making a contact. It is most unlikely that he will be able to narrow things down immediately to the extent where a contact is likely, and the initial use of the troops may well be restricted to getting more information by observation or by looking for tracks in a particular bit of country for example. Alternatively, the troops might be used to frighten the enemy away from one of his possible refuge areas, should he be using it, so as to make it more likely that he will be in one of the other ones. There are endless variations on this theme. Having got more information from the use of the troops, further deductions can be made leading to further use of the troops. Perhaps, this time, enough information will be collected to make a contact with an enemy group possible. Even if the contact is unsuccessful it should help to confirm the validity of the information already obtained and the deductions drawn from it. With luck far more information will result, especially if a prisoner is taken or if letters or equipment fall into the hands of the soldiers. This extra information then has to be added to what is already known, and further deductions made leading to further use of troops and a further contact. The whole process is a chain reaction which gradually builds up and snowballs so that after a time it should be possible to bring the enemy to action under favourable circumstances, and at the same time uncover and root out his support within the population. Unfortunately it is easy for the chain to be broken for one reason or another and when this happens it is necessary to start again from the beginning.

Described in this way the process may seem very long-winded but it is infinitely more effective than employing troops endlessly on a hit or miss basis if the enemy is thinly spread over an area affording good cover. If the enemy is present in greater strength it is still important to use a method which ensures that his organization is systematically uncovered and destroyed, because although if there are enough enemy around the random approach may result in some casualties being inflicted on him, his organization will not be rooted out and the casualties will merely be replaced. Furthermore, if the enemy is operating in such large numbers that it is reasonably easy to find him, it almost certainly means that the situation is deteriorating, and that the transition from guerilla warfare to open warfare is becoming increasingly imminent. If this is the situation it is all the more important that the forces of the government should be employed selectively and to the very best advantage.

Clearly, if a commander is going to operate in this way it means that he must be prepared to devote an immense amount of his time, thought, and energy to handling information. From a superficial point of view this might seem to involve him in devoting too much of his effort in this way, compared to the other functions which he is obliged to carry out. But the point has already been made that the main problem in fighting insurgents lies in finding them, and it could be said that the process of developing information in the way described constitutes the basic tactical function of counter-insurgency operations. It is therefore right that a tactical commander should concentrate on it, and in any case the business cannot be delegated, because the process of absorbing the background information and making plans for the use of the troops based on it is inseparable from the function of command. Either the commander does it or it is not done at all.

In fact this concept is not as odd as it may appear at first sight, and it does at least bear comparison with the procedures used in conventional forms of warfare. For example, a company commander planning an attack gets some information from outside his own resources such as reports from battalion headquarters of information discovered by aerial reconnaissance, wireless intercepts, armoured cars and patrols sent out by other battalions along the front. This equates with the background information provided by the Intelligence Organization or from other sources. In conventional war the company commander than adds to the information he has been given, by looking through his binoculars and by sending out patrols and observation posts from his own resources. This could be said to equate to the use of troops in counter-insurgency operations for gathering additional information and for checking deductions as described. Not until all this had been done would the company commander make his plan, and even then it might be necessary for him to carry out a preliminary assault to gain observation over the area of the final objective. It would be useless for the company commander to blame the failure of his attack on the shortage of information originally given to him if he had made no attempt to supplement it from his own resources. If he adopted a straightforward hit or miss approach and took no trouble to get the sort of information he required in order to make his plan of attack, he would be regarded as a failure.

In most counter-insurgency campaigns the main burden for developing background information falls on the normal military

units and has to be carried out by men using the skills and equipment available to them in the ordinary course of events. In some
cases, however, groups are formed designed to develop information by using special skills and equipment or by exploiting the
characteristics of special people such as captured insurgents.
These groups usually work in close conjunction with the Intelligence Organization or they may be formed within it. Many different ways of developing background information have been devised
by such people ranging from the very simple to the elaborately
planned long-term operation lasting for weeks or months and involving the use of specially raised and trained men. An example of
a simple Special Operation would be the cordonning of a village
and the examination of the occupants by a row of informers concealed in hoods, who are thus able to denounce any enemy present
without the risk of detection. In this case background information
in the form of the name of the informer is developed into contact
information by putting him into a position from which he can
identify and cause to be arrested members of the insurgent
organization. This technique has often been successfully used for
rounding up members of supporters' organizations and of armed
groups who happened to be visiting them. A more elaborate operation might involve the building up of a pseudo-gang from captured insurgents and the cultivation by them of a local supporters'
committee in a particular area designed ultimately to put the
pseudo-gang into touch with a genuine enemy group.

In the last war the Germans repeatedly made use of special
operations against partisans in the occupied countries,[1] and
Filipinos used pseudo-gangs against the Huk guerillas to good
effect.[2] General Massu used turncoats in Algiers,[3] and the British
used captured insurgents extensively in Kenya[4] and to a lesser
extent in Cyprus.[5] There is some evidence to the effect that
pseudo-gangs of ultra-militant black nationalists are operating
now in the United States.[6]

Before leaving the subject it is worth mentioning that under

[1] F. O. MIKSCHE, op. cit., p.159.
[2] VALERIANO and BOHANNAN, op. cit., pp.143–146.
[3] EDGAR O'BALLANCE, *The Algerian Insurrection*, p.80.
[4] FRED MAJDALANEY, op. cit., pp.217–218.
[5] GENERAL GRIVAS, *Memoirs*, edited by Charles Foley, p.105.
[6] MARTIN OPPENHEIMER, op. cit., p.62.

certain unusual conditions a whole campaign might be conducted on the basis of a Special Operation. For example, if a campaign of subversion was being waged against the government of a friendly but primitive country, and if an assessment was made to the effect that it was only necessary to destroy the enemy leadership for the whole campaign to collapse, it might be felt that a Special Operation based on the principles of deep penetration and the use of captured or pseudo-insurgents would afford the best chance of success, providing that a reasonable intelligence organization existed on which to base it. Obviously a system which depends on developing information cannot work if there is absolutely no information to develop, and it is unfortunately true that in the sort of place where a Special Operation could so easily be decisive, there is often no Intelligence Organization at all. Under such circumstances the Special Force would have to build up its own intelligence organization which would take a long time. None the less the use of a Special Operation offers many advantages over the employment of conventional troops. For one thing it is likely to be more economic because most of the work is done by the local inhabitants instead of by the higher paid members of the regular army, and in any case the total number involved is likely to be far smaller. Also there is no requirement for expensive command arrangements or logistic backing. Another important advantage is that Special Forces can operate in a more unobtrusive way than regular troops, and this could be an important factor if the insurgents were trying to exploit world opinion against their government. It is admittedly unlikely that the opportunity for conducting a campaign in this way would occur very frequently, but the army should be capable of working in this way should the need arise.

In preparing the army to fight insurgents, the basic idea of collecting background information and developing it into contact information is of the greatest importance because it is applicable to any counter-insurgency situation. It could perhaps be better described as a method of approach or as a framework into which particular techniques can be fitted rather than as an idea in its own right. The techniques themselves, that is to say the various methods of patrolling, laying ambushes, cordonning villages and controlling food supplies for example, are also important and must be learnt but only a few of them are likely to be applicable in the circumstances of any particular campaign and then they probably require adapting. The value of the approach based on developing information, is that it enables a commander who understands it, to select, adapt and invent particular methods and techniques, suitable to the prevailing circumstances. In order to illustrate the idea as vividly as possible a scenario will be used in which, step by step, the procedure is explained against a fictitious background. Although the background is fictitious, the methods have all been used in one area or another as indicated in the text or in the footnotes. At this point it is only necessary to stress that the purpose of the scenario is to illustrate the idea and not to discuss the specific methods or techniques described.

The Setting
For the purpose of this scenario it is to be assumed that a subversive campaign is in progress in a heavily wooded country such as Malaya or Norway. The enemy is communist but although there are supporters' groups in many of the villages, the population as a whole has not yet been persuaded or coerced into providing whole-hearted backing for the insurgents. At the same time the people are not convinced that the government is going to win through, and are too frightened to offer much practical assistance to it. Most of the senior communist cells directing the campaign live in the forests because they do not consider it safe to live in the villages. In the forests they can hold meetings, with less fear of interruption, and they can communicate with each other and with subordinate cells and supporters' groups living in the villages much better than they could if they were actually resident among the population. Also in the forests are some armed enemy groups whose principal business is to terrorize pockets of loyalists among the civilian population, but who have become increasingly used for drawing attention to the campaign by attacking police posts and by causing damage to government property. In the capital city of the country there is a separate chain of cells and supporters' groups together with a number of armed sections. The city communists are virtually an independent organization although nominally under the command of the supreme committee in the forest. Their own offensive activities are not very important in relation to the damage done, but they are highly significant in

102

terms of their propaganda value, and they also provide supplies, money and recruits for the forest group.

From the government point of view the country is divided into counties and districts which roughly approximate in size to the counties and rural districts of England although the population density is lower because so much of the land is covered in forest. The capital is administered by a city council and is subdivided into wards. Up to the outbreak of the trouble the country was run by elected county and district committees with elected chairmen and members but then emergency legislation was passed which resulted in most of the effective power being transferred to the clerks of the various councils assisted by the local police chief. This arrangement was made possible by the fact that police boundaries coincided with those of the local administration. At the same time steps were taken to build up an intelligence organization within the police which included reinforcement by military officers. Despite these arrangements the situation continued to deteriorate rapidly. At the time in question the police who had been grossly over-worked for months, were known to have been penetrated by communists to some extent. The Clerks of the various councils were thoroughly jittery and several had recently been assassinated. For these reasons the decision was taken by the government to use the army.

The events covered in this illustration should be understood as taking place in one of the districts of the country. This district was part of a very mountainous and heavily wooded county which was therefore particularly valuable to the insurgents as it might easily have provided them with a base area in due course. The whole of the western half of this district consisted of a vast forest and although the rest of it was mainly agricultural there were several other wooded areas in which insurgents could have hidden with reasonable safety for several days at a time. From North to South the District was twelve miles long and from East to West the distance was about eight miles. It contained one small town and five villages in which most of the people lived, but there were a number of odd houses and cottages dotted around outside the villages.

Background Information

Although the government were very short of troops they decided

103

to allocate a battalion to the County mentioned above because of the danger that it might otherwise be developed by the enemy into a base area. The battalion was made available only at the expense of a number of less vulnerable counties which had to be left altogether for the time being. They would have to be dealt with later on, either by newly raised troops or by a redeployment once the original areas tackled had been set in order. Within the County the battalion was split in such a way that one company was available for the District concerned in this illustration. The techniques used in this part of the scenario bear a resemblance to those used by the French in Algeria.[1] A diagrammatic representation of the District is shown in Figure 3.

On arrival in the District the company commander found that the Clerk of the Council had no clear idea of how to deal with the threat. The Chief Inspector of Police was very worried, having lost control of the situation beyond the confines of the one small town which housed the District headquarters. There had formerly been police stations in each of the villages but the insurgents had recently overrun those in villages A and C killing the occupants. The policemen in village B were withdrawn and the post locked up before it could be destroyed. In villages D and E the police stations were still occupied but it appeared that the policemen there were either totally inactive or possibly working on behalf of the insurgents. An intelligence section had recently been established in the police district headquarters but so far it only consisted of one inspector and a constable; a junior military officer was expected to arrive shortly to reinforce it. No one had any knowledge of the insurgents' organization but there were rumours that communist councils were virtually ruling the villages. There was clearly at least one armed group operating in the area judging by the attacks on the police stations, but there may have been many more. Whether they were living in the forest or whether they were living at home and assembling when required was not known. The people were not prepared to say anything at all to anyone connected with the government.

At the first meeting between the clerk of the council, the company commander and the chief inspector of police, an outline plan of campaign was discussed. One possible course of action was to concentrate on destroying the enemy armed groups first, on the

[1] JOHN MCCUEN, op. cit., pp.225–235.

FIGURE 3

DIAGRAMMATIC REPRESENTATION OF DISTRICT

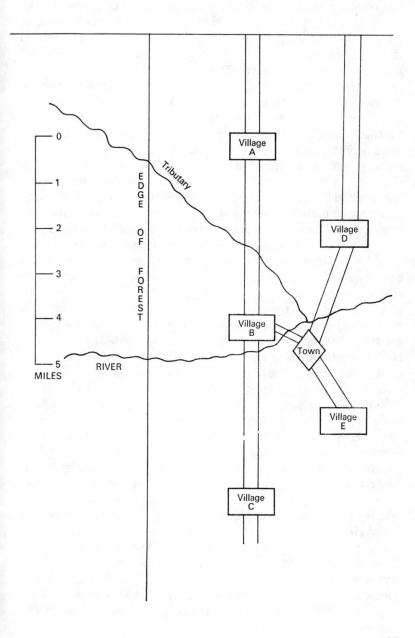

grounds that while they were in existence the population would be too frightened to help the government in any way. In the absence of any specific information this would have to be done by putting the company into a base camp at some strategic spot and then sending out patrols to try and get a lead on to the enemy. It might in fact have been possible to set up two separate bases by splitting the company which would have doubled the chances of success. A plan made along these lines with suitable embellishments would almost certainly have been adopted by a committee inexperienced in counter-insurgency operations but it would have been unlikely to succeed because the soldiers would have been operating on a hit or miss basis, and the communist cells in the villages would have continued to control the area. Even if contact had been made with one of the armed insurgent bands any casualties inflicted would quickly have been made good by recruits from the population controlled by the committees. In any case the enemy operating on good information provided by the population would have been well placed to ambush weak detachments of government forces at the same time avoiding any strong forces sent against them.

A second course was therefore considered which was based on regaining a measure of control over the population, with a view to getting hold of background information which could subsequently be developed into the sort of information necessary for making contact with the armed groups. This plan involved sending a team into each village to work in close contact with the village (or parish) council, some or all of whom were almost sure to need replacing. The teams would have to be led by suitable men from either the army, the police or the administration because no one service would have been able to find enough good men to lead all the teams. The teams would also have to be self-supporting and capable of defending themselves. In addition a good proportion of the army company would have to be held in reserve to go to the rescue of beleaguered loyalists, follow up opportunity targets or guard vital points or important people. When the plan was looked at in detail it became clear that there would not be enough men to form five teams so it was decided to start by raising three whose task would be to restore the situation in villages A, B and C. The teams would be led by one of the military platoon commanders, a police sergeant and the District Agricultural Officer respectively.

Each would consist of a section of soldiers and one policeman, there being insufficient police available to re-occupy the police stations and man them. The sections would be found by breaking up the platoon whose commander was going to be used as a team leader. The other two platoons would be based near the town and would send patrols to visit the police remaining in villages D and E, provide guards if necessary and hold men at short notice to follow up leads and carry out such other operations as might be required in support of the teams. It was realized that it might soon be necessary to concentrate into the villages those people living in scattered houses and cottages around the countryside, both to protect them from intimidation and to prevent them affording succour to the enemy, but it was accepted that there were as yet insufficient resources to enable a programme of that nature to be implemented. It was also agreed that the central government should be asked to provide psychological warfare facilities for disseminating information but it was discovered that none were available.

In addition to the team leader, the team which went to village A consisted of seven soldiers and a police constable who had served in the village some years previously but who had been working in District headquarters since then. On arrival they set about restoring the damaged police station and put it into a state of defence, surrounding it with trenches, earthworks and wire. The team leader approached the chairman of the village council who was the owner of the village shop, but although he found him polite, he got no help from him. He was told that there were no active communists in the village and that no one knew anything about the armed group which had overrun the police station. Other prominent persons in the village were approached and gave similar accounts of the situation but they were not all as polite as the chairman.

Faced with a complete blank the team leader decided to make a census of the village. Starting at one end he went through one household after another. In every house he took aside each member of the household in turn and questioned him briefly and in private about his work, the whereabouts of close relations and other domestic matters. He kept a record of these talks and at the end of each one gave the individual concerned a card with his or her name, address and particulars on it, to be carried and produced when required. He also compared information given by different members of the family and if there was any discrepancy regarding particulars of an absent member or about a relation living outside the district, he went back and probed around more carefully. Occasionally he rang up the police in a different part of the country and asked them to check whether a person whom he was told had been living at a certain address was in fact living there. The policeman on the team was useful for filling in gaps in the information provided by some of the families and for contacting police stations in different parts of the country. The team leader drove himself hard, restricting time spent in eating and sleeping to a minimum and he completed the job in about ten days. At the end he still knew nothing about the communist organization or about their armed groups, but he knew a lot more about the village and its people so that he was able to recognize and greet by name individuals he met as he moved around. In this way he felt that he was building up a store of good will which might prove useful in the months ahead.

During the ten days during which the census was being carried out the seven men of the section under their corporal continued to improve the defences of the police station. This work, allied to the fact that they had to keep two sentries on watch throughout the twenty-four hours, did not leave much time for other activities, but they did send out patrols by day into the immediately surrounding areas and they also sent men to the village school every now and then to help organize sporting and physical training exercises. This was very popular with the children but oddly enough the school-master did not appear to be as enthusiastic as he might have been. Once, at night, the soldiers had a practice 'stand to' which included calling on company headquarters near the town to send out a relief force.

The team leader was still in no position to take any sort of action against the communists but he was ready to go a stage further by using the background information which he had obtained to get more information. To this end he decided to narrow down the field of investigation into certain specific channels from which he hoped to produce concrete results. He decided to approach the problem along four main routes. First he would work out which people were best placed to help the communists by virtue of the work which they were doing. For example a man whose business involved felling trees in the forest and transporting them to a mill outside would be in a better position to take food to armed groups hiding there than a man whose job kept him within the village itself. Second, he would enquire carefully into the financial situation of a selected group of villagers to see whether they appeared to be spending or saving all the money they were earning. If there were grave discrepancies between one person and the next it might indicate payment into party funds. Third he would investigate the past political affiliations of a few of the better placed people in the village to see whether they were likely to be in sympathy with the aims of the insurgents. Fourth he would make a complete list of all the livestock owned by members of the village and watch how it was disposed of. His aim at this juncture was not to prevent food reaching the enemy but to try and identify people who were supplying it.

It cannot be said that any spectacular discoveries resulted from this study. Although the team leader did get some general indication of the people who were best placed to help the communists, he did not find anyone actually doing so. When it came to money he found that in each case investigated the combination of a man's saving and expenditure was less than he thought that it should be on the basis of likely income, but he could not be sure whether this was because everyone was contributing to the enemy funds or whether he had merely made his assessments incorrectly. In the case of one farmer he noticed that his turnover appeared to be much larger than he expected. His political investigations revealed that several men in the village had formerly been members of the communist party in the days before it was banned. When he talked to them they all said that they were totally opposed to present communist policies and seemed quite willing to discuss some of the ideas and characteristics of their former leaders. He made notes of all they said but found that none of it was relevant to happenings in village A. He reported it to District headquarters for what it was worth. In watching the cattle and sheep he noticed three cases in a fortnight of animals disappearing: once the owner complained to the policeman that he had lost a bullock but on the other two occasions no mention was made of the matter.

After studying the results of all the information which he had collected, the team leader thought that he would try some shock treatment. He had come to the conclusion that the farmer whose financial turnover was higher than it should have been, might be the local treasurer collecting and passing on money for the communists. He therefore decided to interrogate four of the people who from his investigations seemed least likely to be supporting the communists willingly with a view to seeing whether any of them would confirm the farmer's position as treasurer. He arranged for the policeman to arrest all four one day for various trivial offences and bring them to the police station to be charged. Whilst they were in the police station he talked to each one separately and privately, warning him that although he was only being charged with a minor offence he happened to know that he was contributing to the enemy's funds for which he could be sent into a detention centre. Although one flatly denied this, the others said that everyone in the village was obliged to contribute. Playing on the evident annoyance felt by one of the men for having to give money to the communists the team leader got him talking to some extent. At first he was very nervous of talking at all, but the team leader pointed out that he could easily deny having said anything, and that as four had been arrested together no one outside would ever know which of them had given information. In fact no one need know that any information had been given. Eventually the arrested man agreed that he paid his contribution to the farmer but he insisted that he had no idea who the other members of the communist leadership were in the village. In due course all four arrested men were taken to court and fined a small amount for the offence for which they had ostensibly been arrested.

From then onwards the team leader naturally watched the farmer like a hawk, and by process of deduction and investigation he soon had a very shrewd idea of the identity of two or three members of the communist committee which included the schoolmaster. Meanwhile he passed back to the police inspector running the intelligence section at District headquarters, all the information which he had collected from the start and discussed with the company commander the desirability of arresting the principal communists whose identity he had discovered. The company commander told him that a similar pattern was emerging in villages B and C and that the District Committee, i.e. himself, the clerk of the council and the chief inspector of police, had decided to leave things for a bit longer and make a clean sweep of all three communist village committees at the same time once they felt that they knew enough to catch a reasonable proportion of the principal personalities involved.

In the event the communists forestalled the District Committee. Somehow they discovered the name of a man in village B who had given information to the team leader and one morning he was found dead on a bit of waste land outside the village surrounded by the bodies of his wife and three small children. The communist committees of all three villages had disappeared into the forest during the hours of darkness. In village A the farmer and the school-master both went, together with the owner of the shop. Two nights later the police station was attacked by a large armed group and the team only just succeeded in holding out behind their defences until help sent by the company commander arrived. For the next few weeks a series of incidents occurred throughout the District which kept the company fully occupied: the enemy seemed capable of striking pretty well where they pleased and the company was obliged to react, although it did succeed in inflicting casualties on the communist groups. At one time it looked as if the teams would have to be withdrawn from the villages in order that the third platoon could be reconstituted, and the defence commitment reduced, but this was so obviously what the enemy were striving to achieve that the District Committee decided to hang on for as long as they could. Within the three villages themselves the situation improved immeasurably, and a few people even started giving the team leaders names of communist supporters who were arrested and removed. Most of the people merely became more friendly but did not actually commit themselves too far, in case the communists managed to re-establish control at some later date. After a few weeks the enemy attacks eased off and the situation became more stable. The government were genuinely in control in villages A, B and C but the communists retained their influence in villages D and E and amongst the people living in the country outside the villages.

By this time a great deal of background information had been obtained, not only by the teams in the manner described, but also as a result of the interrogation of arrested communist supporters and from the identification of enemy dead. If there had been any troops available for the purpose, it would now have been possible to start offensive action against enemy armed groups on the basis of developing this background information into contact information, but the District Committee decided to give priority to restoring the situation in those parts of the District where the communists retained their influence. At one moment it seemed possible that crisis conditions in a neighbouring district might necessitate re-deploying the company there in an attempt to stave off impending disaster and if this had happened the communists would have immediately regained control in villages A, B and C, killing all those who had afforded help to the team leaders during the time the teams had been operating. The only thing that would have been salvaged from this wreck would have been the priceless background information which was by that time safely recorded by the intelligence organization. Luckily the company was allowed to remain.

Developing Information

Although the company was not moved at that critical early stage, later developments in the District inevitably reflected the progress of the campaign throughout the country as a whole, and the moment came when the soldiers had to leave for service elsewhere. Luckily by the time this happened the government was in control of all five villages and the teams had already been re-placed by new village leaders supported and protected by locally raised Home Guard detachments. Enough police reinforcements had been received in the district to enable the police stations in villages A, B and C to be re-opened and manned. Most of the outlying population had been resettled in the villages and fairly strict control of livestock and crops had been instituted to prevent food from reaching the enemy's forest groups.

The next part of the illustration concerns the activities of a new military unit trying to contact the enemy. In order to make the problem a difficult one, the setting depicts a situation in which enemy numbers have been considerably reduced compared to the last section of the scenario and it thereby supposes that the government campaign has moved successfully into a later phase of operations. The operational principles illustrated are this time based on the author's experience in Malaya and have not previously been explained in print to the best of his knowledge although they are hinted at by General Clutterbuck in his book on the Malayan Emergency.[1]

[1] RICHARD CLUTTERBUCK, op. cit., p.108.

When the new company commander arrived in the District he found a very different situation to that which had confronted his predecessor at the beginning of the campaign. In District police headquarters there was now a combined operations room, the walls of which were covered by maps showing the locations of police and Home Guard units and recording the site of all incidents which had taken place since the start of the trouble. These map displays were supplemented by files of historical data. The intelligence section had its offices nearby and the inspector in charge, assisted as he was by an army officer, was ready to brief the company commander on his arrival. He started by explaining that the enemy leadership consisted of a District Committee and two Branch Committees, one of which was responsible for the country North of the river and one for the country South of it. The communist District Committee had been living in the forest since the party had been prescribed two years before the start of the campaign, but the two Branch Committees had been set up, also in the forest, soon after the village committees had been forced out of the villages as a result of the activities of the original teams. The Inspector explained that since the departure of the first company the situation had deteriorated to some extent and that the communists had re-established supporters' cells on a small scale in some of the villages and were even suspected of having some covert supporters in the Home Guard. Members of the Branch Committees were known to leave the forest from time to time to visit their supporters in the villages with a view to extending their influence and collecting supplies. On these occasions they were escorted by one or two armed bodyguards. Also living in the forest, but physically separated from the committees, was a well-armed insurgent platoon whose job was to carry out tasks on the order of the District Committee in support of the Branches. If a really large force was required extra platoons could be drafted into the area on the orders of the communist County Committee which itself lived in the forested part of the District for some of the time. In addition to this information the intelligence section had identified all the members of the District and Branch Committees and had identified most of the members of the supporting platoon. It had records of some of their weapons and it knew which villages were most heavily indoctrinated and therefore most likely to be supporting the armed groups. Sometimes the intelligence section even got reports of communist committee members visiting their supporters which were only a few hours old, but usually there was a longer time lap. Over the past few months they had on two occasions received advanced information about meetings, or about visits from members of the platoon itself to a village to pick up supplies.

At his first meeting with the clerk of the council and the chief inspector of police the company commander discussed a plan for destroying the enemy in the forest. All three of them agreed that it was desirable to tighten up on measures designed to prevent food reaching the enemy, but both the clerk of the council and the chief inspector said that they could not spare any additional men or set up extra control points. If more was to be done the company commander would have to use one of his three platoons in this way. For the rest the chief inspector advised the company commander to keep a patrol with a tracker dog at a few minutes notice to rush, day or night, to the scene of any reported sighting. He also said that the best chance of making effective contacts was to train the men in ambush drills and be ready to act on such pinpoint information as the intelligence section might produce over the months. He recommended that the rest of the company should be divided into patrols and sent to search the forest, map square by map square, on a hit or miss basis. He said that if only the District Intelligence Section was worth its salt a programme of this sort would be bound to succeed after a short time. The chief inspector's advice closely followed the lines of most company programmes but it would not have worked for the simple reason that the intelligence section could not possibly supply enough of the sort of information which the chief inspector had in mind, and a map square by map square search of forty-eight square miles of thick forest could hardly be expected to produce more than a chance contact or two within the lifetime of any of those present.

Fortunately the company commander realized that this was the case and that he would have to work out a better system along which to conduct his operations. He was not content merely to blame the intelligence organization for lack of the right sort of information although his colonel seemed ready enough to do so. The company commander had been given, or had access to, a lot of good background information and he knew about the theory of developing it into contact information. His problem was to apply the theory in such a way as to bring about the destruction of the four armed groups living in the District, that is to say the communist District Committee with its guards and escorts, the two Branch Committees with their guards and escorts, and the insurgent platoon supporting them. He discussed the question of priority with the clerk of the council and the chief inspector of police and they all agreed that the insurgent platoon was not doing enough by way of terrorizing the population as a whole, to make its destruction a matter of urgency and that priority should therefore be given to the committee groups. Obviously it would be desirable to destroy the District Committee first, but they all realized that it was less vulnerable than the other two because its members never left the forest to contact the cells outside, either for the purpose of directing their efforts or for collecting food which came to it through the Branch in whose area it happened to be living at the time. The company commander therefore decided to concentrate his efforts against the Branch Committee groups, giving equal priority to both initially, but possibly concentrating on one of them later on if the information which he developed led him in one direction more than the other. Naturally he would take advantage of any chance opportunity which might bring him into contact with the communist District Committee or the insurgent platoon.

In practical terms the company commander's job was to narrow down the whereabouts of the enemy both in terms of space and in terms of time to the extent where he could put his men into contact with them. As a start he asked the intelligence section for a list of the names of the insurgents in the two groups. There was no doubt about the identity of the committee members themselves but the inspector pointed out that the guards and escorts sometimes changed round with members of the insurgent platoon, so the company commander was given their names as well. He then studied the lists with great care asking for as much detail as possible about each individual such as the area in which he had lived before going to the forest, his occupation at that time, and the present whereabouts of his family. The intelligence section could only give him some of this information but he managed to supplement it by visiting the villages and talking to the local Home Guards or policemen about insurgents who had originated in the area. Sometimes he drew blank but sometimes he found people who had known one of the men on his list and then he might gain some interesting piece of knowledge or at least an insight into the man's character and likely way of reacting to particular situations. After a time a pattern emerged which showed that the Northern Branch had more contacts with village A than with B or D and in particular that the leader of the committee and two of the members had originated from it. The third member had come from village D. Strangely enough this Branch had no direct connection with village B but one of the guards had formerly lived on a farm near the forest edge and his family had been resettled in village B. The pattern concerning the Southern Branch was less clear cut, members being about equally drawn from villages C and E.

The next subject which the company commander investigated was the degree of support which the communists had in the villages and in the town. In this direction the intelligence section was less sure of the facts, but said that cells were known to exist in villages A and C and it was also known that the Northern Branch were trying hard to re-establish one in village B. There was no evidence about the situation in the town or in villages D and E. Cells had formerly existed in all these places but they had been rooted out some weeks earlier and no positive information about successor organizations had been received. On the other hand the inspector made it clear that his coverage of the area away from the forest was far from satisfactory.

The company commander then turned his attention towards the forest to see whether there were any considerations which indicated that some parts of it were more likely to be in use as living areas than others. Obviously the Branches would have liked to live as near to the forest edge as was consistent with their safety because they had to visit the country outside it to get supplies and to contact supporters. The safety factor would have varied in accordance with a number of other factors such as the thickness of the cover in any particular area, and the company commander sent out a number of short reconnaissance patrols in order to get information on the subject. He also had a good look around for himself in order to get the feel of his bit of forest and he went through the records of all past contacts to see where they had been made and what had been found in each case. This at least gave him some idea of the sort of places which had been used as camp sites in the past. He next related this to the time of year in which the contact had taken place to see whether there was any seasonal influence discernible, which could have resulted from the level of water in the river and streams for example, or even from an indirect cause such as the presence of loggers in a particular part of the forest at a particular time of the year.

The company commander found that one of the most profitable areas of research lay in the informal talks which he had with captured insurgents. He arranged to have two of these people attached to the company, both of whom had been members of insurgent groups operating in the District but who had subsequently proved their willingness to co-operate with the government. Whenever he drove around the area he took one of them with him, both to point out places of interest and to talk about life in the gangs. Both men had been captured some time earlier and they had no up-to-date information about enemy locations or intentions, but they did reveal a lot about their pattern of life. For example they knew all about the routine followed by the communist groups and they could say whether movement in any given area was more likely to take place by night or by day; this matter being largely dependent on the thickness of cover in the place concerned. One of them explained that in his time, when a Branch Committee member wanted to contact supporters in village D, he would stay with his escort in a hide near the village for a night or two, but if he wanted to contact friends in village B he would do it direct from the forest camp. They also pointed out to the company commander spots which enemy groups usually used for crossing the main road or the river, and which tracks and paths were particularly favoured by one leader and which by another.

After a time the company commander realized that he was making more headway in the North than in the South so he decided to concentrate his efforts on the Branch which was operating there. On several occasions individual insurgents were reported as being seen talking to villagers during the morning just outside the forest in the area of village B which probably meant that they were busy building up support there. The company commander had already worked out that the Northern Branch usually lived in one of four forest areas shown on Figure 4 as Living Areas, W, X, Y and Z. The sightings made it almost certain that they were in X or Y. He had also worked out that these insurgents

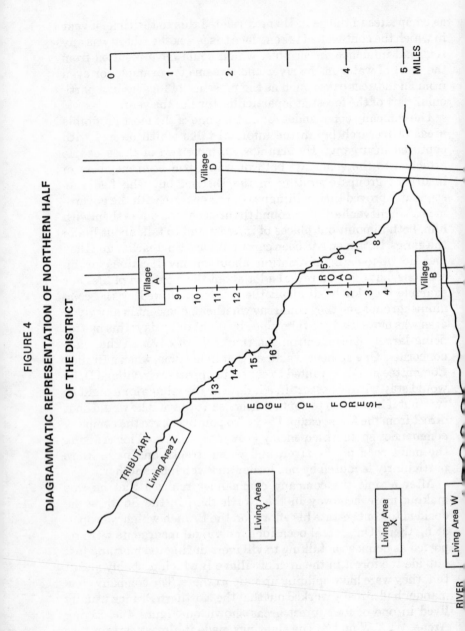

FIGURE 4

DIAGRAMMATIC REPRESENTATION OF NORTHERN HALF
OF THE DISTRICT

118

would make a sortie for food to village A or village D during the course of the coming month. If the sortie was to be to village A it would probably take place between the sixth and the twelfth because at that time there would be a moon in the early part of the night which would enable them to travel through the forest by its light but which would give them darkness for the actual approach to the village. If the sortie was to be to village D it would probably take place two weeks later because the initial approach would be on one night and the pick-up and return on the next night. In this case a dark early part to the night would be required.

The Chief Inspector of Police and the Clerk of the Council both put pressure on the company commander to follow up reports of sightings with a tracker combat team and to send patrols into the two areas of forest in which they thought that the Branch might be living, but he resisted strongly. He knew that the sighting reports were all much too old to give such a team any reasonable chance of success and that if he did react to the reports he would merely alert the enemy to the fact that informers were operating against them in the area. He also refused to send patrols into Living Areas X and Y because although they represented much smaller target areas than the forest as a whole, they were still far too big to ensure a satisfactory encounter. Patrols operating in them might conceivably bump into an enemy camp, but if one did the chances would be heavily weighted in favour of the insurgents' escape. A far more likely result would be that the enemy would move to one of the other living areas where it would be more difficult for the soldiers to keep track of them. In any case, having got so far with his calculations the company commander was not going to spring the trap until he was sure of some reasonably substantial result.

His next move therefore was to plot on the map all the likely places where the insurgents would cross the road, and all the likely places where they would cross the tributary if they were moving from Living Area X to village A. His calculations, which were largely based on the information given him by his captured insurgents, indicated that there were eight possible ambush points shown on Figure 4 by the figures 1 to 8. He then repeated the process on the basis of a journey from Living Area Y to village A and produced another eight possible ambush points indicated on Figure 4 by the figures 9–16. Thus in order to be

119

sure of making a contact when the insurgents made their sortie he would need to have sixteen ambushes in position for the six nights when the moon was right. This was just about twice what he could provide using the maximum resources of his company. Although he had narrowed down the possibilities for a contact both in terms of space and in terms of time, he had not done so sufficiently to ensure success. For a moment he toyed with the idea of selecting the eight most likely points out of the sixteen but he decided that there must be a better system than that. He then wondered whether he could make certain of a contact by putting his ambushes in a tight ring around the village but abandoned the idea because undoubtedly one of them would be discovered by a villager and word would be passed to the insurgents who would promptly take fright and cancel the whole project. He therefore decided to utilize the period before the sixth of the following month to discover whether the insurgents were in fact in Living Area X or Living Area Y. If successful he could then cover all contingencies with eight ambushes.

The company commander set about doing this by sending a number of tracker reconnaissance patrols into the forest as soon as he heard that there had been another sighting near village B. The patrol commanders were told to keep well away from the forest edge so that villagers should not connect them with the sighting; they were also told to keep out of Living Areas X and Y so as not to disturb the insurgents there. They were instructed to search the forest between these two living areas and village B to see whether they could find any tracks or indications of movement. After a few days they found evidence which made it clear that the enemy were in Living Area Y. The company commander had now enough background information and had developed it by his own resources to make it possible for him to put his men into contact with the enemy under favourable circumstances.

Unfortunately the eight ambushes drew blank despite the fact that they stayed in their positions between the sixth and the twelfth of the month. The company commander was disappointed but re-plotted the whole operation on the basis of a move between Living Area Y and village D two weeks later. This time he was successful and five insurgents walked into one of the ambushes, three of whom were killed. This was a great triumph because, apart from anything else, the very presence of the enemy at that spot and at that time made it plain that the company commander's calculations had been based on correct assumptions. Furthermore providing that the two survivors returned to the other members of the Branch in the forest quickly, there was no reason why they should move to another area because they had no reason to think that the government forces would know where they were. The insurgents would regard a contact so far from both village and living area as no more than bad luck.

One of the dead men was identified as the second in command of the Branch: the other two were rank and file escorts. Unfortunately there were no live prisoners to interrogate and no useful documents were found on the bodies. It looked as if the company commander would have to start again and try and work out when next a sortie for food was likely to take place, and in fact he did start work along these lines. But there was one worth-while clue which arose out of the ambush. Amongst several sacks of food taken from the bodies at the scene of the ambush was one which contained some fruit which was seldom grown in the area. The company commander and the inspector of the intelligence section discussed this and came to the conclusion that if a farm growing such fruit could be found near village D, there would be a chance of getting some useful information out of the owner. The company commander therefore went to village D and told the commander of the local Home Guard that he had reason to think that some insurgents were hiding near the village. He suggested that the Home Guard and the army should do a joint sweep through the area and look for them. In fact the Home Guard in that place were heavily penetrated by the communists but knowing that none of their friends from the forest were in the area they agreed to take part and the sweep was carried out with great enthusiasm by all concerned. Naturally no insurgents were found but the soldiers who had been carefully briefed in advance discovered two farms on which the fruit in question was being grown. Next evening the inspector of the intelligence section had long private talks with the owners as a result of which he discovered that one of them had a brother-in-law who was with the insurgents and that he provided the fruit on that account. He also discovered that the brother-in-law was responsible for organizing food supplies for the Branch and that he stayed at the farm whenever he came to arrange the sorties. The inspector worked on the farmer and his wife for some time and eventually persuaded them to betray the man in return for a promise that he would be leniently treated and that they would all be rewarded. He also made it clear that the forces of the government were hot on the trail of the insurgents in the area and that the only ones likely to survive were those that were captured before they could be killed.

Some days later the inspector heard that the brother-in-law was due to pay a visit to village D the following evening. When this insurgent arrived he was met by the company commander and the inspector who immediately started trying to persuade him to lead some soldiers to the forest camp where his friends were living. There was no time to lose. The man was due to return in two days' time and if he failed to arrive punctually the insurgents would immediately move to another camp on the assumption that he might have been captured and forced into betraying their position. The farmer and his wife joined the company commander and the inspector in trying to persuade the prisoner to co-operate. Eventually they succeeded and by dawn he was leading a strong army patrol into the forest. This time the company commander had no fears about frightening the insurgents away from an area where he could keep track of them into one where he could not. The soldiers were acting on first class, up-to-date information and there was no danger of them blundering into the enemy camp with all the advantages on the side of the insurgents. They had a guide who knew exactly how to avoid the sentry and with a bit of luck they would kill or capture the whole group. The company commander, helped by the intelligence inspector, had succeeded in developing the information gained in the ambush, i.e. the fruit, by further action, i.e. the sweep which discovered where the fruit had come from, into contact information, i.e. the prisoner who would lead them to the enemy's camp. For the purpose of the illustration the process may be regarded as complete and it can be assumed that the patrol was successful. In practice it might not have worked, or it might have been only partially successful in which case the procedure would have to be developed through some extra stages.

Special Operations

The next part of the scenario deals with the business of using special people and specialized techniques to develop background information into contact information. It is based on the author's experience in Kenya.[1]

While the company was engaged in the North of the District the Intelligence Section was busy building up its coverage in the South. After some initial set-backs the inspector managed to get an informer into village C who provided the names of several prominent communist supporters in the area. This was the first good information to have come out of the Southern part of the District for some time so it was decided to take no action against the people named. Instead the intelligence section was invited to watch events in the hope that it might get some further leads on the communist organization in that part of the District. At the time it seemed more important to get a line on the enemy's activities than to curtail them.

During the past few months, throughout the country, the military officers who had been drafted into the intelligence organization had amongst other duties been experimenting in the use of captured insurgents against their former comrades. As part of these experiments the officer in the District covered in this scenario had collected a team of such people together which he kept in a post a few miles outside the town. The intelligence section now decided to use these people against the communist Branch operating South of the river, in an attempt to exploit the information provided by the informer. As a first step a man was selected from the team and provided with a detailed story. In outline the story was that he had been a member of an insurgent platoon which had been engaged in a battle with the forces of the government a few miles outside the forest, in a District which was some miles away. In the story four members of the platoon had been detached from the main body and had hidden in a farm yard, as the forces of the government pursued the rest of the platoon. The whole area seemed to be swarming with troops so the four split up. Soon afterwards this man concealed himself in the back of a lorry and got taken out of the area. Some hours later the driver stopped outside village C for some reason or other and the man jumped out and hid by the side of the road. The man was rehearsed with great care. He was made to learn the personal details of every member of the platoon from which he had allegedly come and these included the particulars of an individual who

[1] F. E. KITSON, op. cit.

123

had originated from village C and who had been killed some weeks earlier. Although it was not usual for men to be drafted into platoons operating away from their own District it had to happen from time to time when heavy casualties were incurred and this was a genuine case which would have been well known to the people of village C. In addition to the story the man was equipped with all the right documents, clothes, and a weapon.

One evening this man was dropped off near village C by the officer from the intelligence organization and told to go to the house of one of the villagers named by the informer as a prominent communist. On arrival he told his story to this villager and asked for help. He said that the insurgent from village C who had been in his platoon up to the time of his death had advised him to contact this man if he ever needed assistance in this part of the country. He said that all he wanted was to be put in touch with the local Branch organization who would know how to get him back to his platoon or who would allocate him to some other unit. The villager was worried but everything about the story was so right that he took the man in and hid him in his house. Later in the evening three members of the Village C Supporters' cell came in and questioned him for some time. Eventually they seemed satisfied and told him to go and hide in a bit of swamp outside the village until they could arrange for him to be put in touch with someone from the forest. They would signal to him to come in again by hanging a red blanket on the washing line which he could see from the swamp where he would be lying up. He slipped out of the village just before dawn and went to the swamp. Later he left his hideout and met the officer from the intelligence section at a pre-arranged rendezvous. In the course of the night he had learned a lot about the communist cell and was able to describe the three members who he had met. The background information provided by the informer combined with the background information which had gone into the cover story had been developed into more information by a special operation but it would be necessary to go a stage further before it could be used to put the government's forces into contact with an armed enemy group.

After hearing the man's story the officer decided to set up two more members of the team and himself to represent the other three members of the platoon who had become separated in the battle. They would join the man in his hideout in the swamp. Then when the supporters in the village sent for him he would explain that there were in fact four escaped members of the platoon and not one. He would say that one of them was a high ranking officer of the organization but that he had not dared mention it at the first meeting because he could not then be sure that the villager and his friends were not planning to betray them. He would say that they had been lying up near, but not at, the place recommended and had watched to see whether any police or soldiers had been tipped off to look there. As none had appeared they concluded that they could trust the village cell and tell them the whole truth about the leader.

At the second meeting which took place three nights later the man told his story. This time there were two extra members of the village cell present together with an insurgent from the Branch in the forest. He was just visiting the village and was going on to the town before returning to the forest but he said that he would be back in a week with four others to pick up some supplies. He said that the man and his three companions could accompany them back to the forest at that time if the leader of the Branch agreed. After the meeting this insurgent went to the hide in the swamp and met the other three. He questioned all of them carefully and appeared satisfied. The original background information could be said to have been developed into contact information because it was now known that a group of insurgents were going to come to a certain place, i.e. the villager's house, at a certain time, i.e. a week later. On the basis of this information it would have been a comparatively easy matter to have destroyed the insurgents.

But the officer was after bigger game. Had he sprung the trap at this stage it is likely that only one Branch member and a few rank and file escorts would be eliminated. He decided to take the matter a stage further and get himself and his three men taken to the Branch headquarters in the forest. Once there he would find out as much as possible about the communist District Committee and the insurgent platoon. He would then choose a favourable moment to turn on his hosts and kill as many of them as he could, making sure that the leaders did not escape. He and his men would take a chance on escaping in the confusion. In this case they would not only have developed background information into contact information but by acting on it themselves at the most favourable possible moment they would reduce to a minimum the chances of the enemy escaping. They would also stand a chance of getting more information about other enemy groups, i.e. the District Committee and the insurgent platoon which might later be developed into contact information.

There is no point in pursuing this part of the scenario any further because the principle of developing background information into contact information using special people and specialized techniques has been sufficiently demonstrated. Suffice it to say that there are innumerable ways in which the principle can be applied under various circumstances and it is up to those involved to invent or adapt such methods of achieving their aims as may be relevant to the situation.

END OF SCENARIO

Although the idea of developing information has been illustrated in relation to events in a rural and forested area, it is equally applicable to other surroundings or to a higher or lower pitch of insurgency. Furthermore the way in which the information is developed does not have to be based on the enemy's supply system as depicted in the scenario. The initial break-in could just as well be made as a result of contacts which an enemy group was making with its supporters in order to disseminate propaganda, or it might come as a result of a personal contact between an insurgent and his family, or it might come in any of a dozen other ways. The situation depicted in the scenario is, of course, greatly simplified by comparison with anything that would be likely to occur in real life. For one thing, in the case of deduction, whenever one factor is discussed there would probably be six or more, and whenever two or three stages are depicted, there might be several times as many. More important still, the scenario does not go into detail about where the troops come from. If they belonged to an ally or to a colonial power, a whole host of complications would arise ranging from language difficulties to reluctance on the part of the population to co-operate with foreigners against their own people. All these problems, and many more besides, are perfectly genuine, but the scenario is designed simply to illustrate the idea of developing information and it has therefore been stripped of anything irrelevant to that purpose. It is certainly not suitable for use as a tactical guide in any actual situation, nor can the situation depicted be regarded as being at all typical.

The business of handling sabotage and terrorism in an urban area deserves special mention because it might appear impossible to track down and destroy very small groups of terrorists hiding in large urban rabbit warrens by means of the system illustrated in the scenario. But urban terrorists like other insurgents suffer from an important weakness in that their actions must be related to a purpose, which in turn involves building up support amongst the population, at the same time causing it to act in accordance with a programme designed to achieve the aims of those running the subversion. There must therefore be a point of contact between the people and the political or controlling wing of the subversive organization on the one hand, and between the controlling wing and the actual terrorists and saboteurs on the other. Using a suitably adapted version of the procedure outlined in the scenario it is possible for the forces of the government to get a lead first into the subversive control organization and then into the terrorist groups themselves.

In order to see how the procedure might work in practice it is necessary to look briefly at the sort of enemy organization which is likely to exist in a large city: this is shown diagrammatically in Figure 5. The organization here depicted is reasonably sophisticated and assumes that an overt political party or movement such as a Civil Rights Association has been partially penetrated by a subversive group, which in turn controls terrorist and sabotage sections, either at the city level as was the case in Nicosia during the EOKA campaign, or at a lower level as in Nairobi during the Mau Mau Emergency. There are innumerable ways in which this pattern might be varied, but in this case it is assumed that only a few of the members of the overt committees are also members of the subversive organization; some of them may not even know of its existence. Subversive organizations usually try to get their members into the most influential positions on the overt committees, but they are not always able to do so.

In order to collect background information the forces of the government must first afford some protection to the uncommitted

FIGURE 5

DIAGRAMMATICAL REPRESENTATION OF A
SUBVERSIVE ORGANIZATION IN AN URBAN SETTING

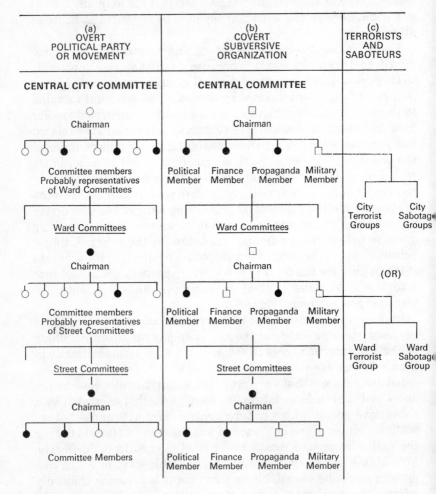

Notes

1. White circle indicates member of overt committee who is NOT a member of the subversive organization.

2. Black circle indicates member of subversive committee who is also a member of an overt committee. Each black dot in column (a) represents the same person as one of the black dots in column (b).

3. White squares represent members of subversive org who are NOT on overt committee.

4. There is no relationship between appointments held and the question of whether a man is a member of an overt committee as well as a covert one.

128

part of the population and then start to organize it. Obviously the methods described in the scenario as being suitable for use in a village in a remote country area, would not be applicable in a city, but a measure of protection might be achieved by setting up small military or police posts to cover each street or groups of streets preferably in positions which provide good observation and fields of fire. Following the procedure used by the French army in Algiers, the policeman or soldier in charge of each of these strong points might then appoint one local inhabitant to be responsible for each street who would be instructed to appoint an individual to be responsible for each block and so on down to one individual responsible for each family. The avowed reason for doing this would be to facilitate requests by the people themselves for help in the event of terrorist activity, intimidation, or any of the other things which they will be giving as their reasons for not being able to help the forces of the government. The men who take up these positions will probably be put forward by the street committee of the party or movement, and they may do their best to mislead the forces of the government, but if the soldier or policeman in charge works along the lines depicted in the scenario a certain amount of background information will be picked up. In particular there should be no difficulty in discovering who the members of the overt committees are, especially as they will be obliged to deal directly with the authorities with regard to any non-violent action which they may wish to initiate as part of the over all campaign.

By entering into discussion with the members of the overt committees when incidents provide a plausible reason for doing so, and by bringing the individuals responsible for streets, blocks, and families into these discussions as required, it should be possible to build up a picture of the relationships which exist between the prominent individuals in the area. From this point it is a short step to gauging which members of the overt committees are also members of the subversive organization. Once suspected, these men can be watched in such a way that some of those who are members of the subversive organization but who are not members of the overt organization will be discovered. So far as the members of high ranking committees are concerned this is work for the intelligence organization but at ground level ordinary soldiers or policemen will have to carry out the task. There are plenty of other ways of getting background information and most soldiers and policemen are quite capable of getting it in a friendly and civilized way, provided that they are told what is required.

When sufficient background information has been built up, policemen or soldiers can act on it in a way designed to get more information, using the techniques most suited to operations in an urban area such as the checking of vehicles and individuals in road blocks or the searching of buildings. These techniques can of course be used at random in an urban setting in the same way that patrols and ambushes can be used on a hit or miss basis in the country, but if background information is available they can be used as part of a chain reaction designed to develop it into contact information. This development can either be done by the local tactical commander using a system of deduction and observation, or it can be done by special units. Special operations are particularly effective in urban areas providing that a satisfactory system for co-ordinating them with the activities of normal government forces can be worked out. This is always a problem, but particularly so in urban areas where troops are inevitably more thickly spread on the ground and more permanently resident than they are in country places.

* * * * * *

If the concept of developing information is accepted, two practical problems have got to be faced when it comes to preparing the army for the task. The first of these problems concerns continuity because the whole core of the business undoubtedly lies in the deduction process which by its very nature takes time and which can only be carried out by the tactical commander. If tactical commanders are changed too frequently no long-term development of information will be attempted and officers will for ever be aiming for quick results in terms of numbers of insurgents killed as opposed to enemy organizations rooted out and destroyed. Furthermore the tactical commander will always be at a disadvantage compared with his opponent who may have been operating in the District for months or even years before he arrived. Inevitably the tactical commanders change whenever units are moved in or out of an area and the problem is therefore one of finding a way of leaving them in an area for long enough. But this is not easy to achieve because there are seldom enough troops to cover the whole country in sufficient strength so some redeployment is usually inevitable in connection with the overall plan of campaign. Furthermore, breaks in continuity occur whenever a unit arrives in or departs from the theatre of operations, and unit tours in theatres of counter-insurgency operations are usually limited to a year and often to a period of a few months. Methods of promoting continuity are mentioned in the next chapter and in Chapter 9 but it is necessary to understand the need for it at this point.

The second point arising out of the development of information principle concerns the way in which the Intelligence Organization works. It has already been mentioned that peace-time intelligence organizations prefer using a few high grade sources to a large number of lower grade ones. But it is evident from the scenario that the system for developing background information only works if there is a lot of it to develop. It is not important that it should be immensely reliable because all that is needed is something on which to build. The military requirement is therefore for an intelligence organization geared to work in a different way to that in which the peace-time organization works. It is up to the army to stress this point in the early stages of the campaign and to see that it gets the sort of information it needs.

In conclusion it must again be stressed that the methods by which commanders can develop background information into contact information in wars of counter-insurgency vary to such an extent between one place and another that they cannot be taught in the same way that the tactics of conventional war can be taught. Commanders can, however, develop a suitable frame of mind which will stand them in good stead anywhere. The process is a sort of game based on intense mental activity allied to a determination to find things out and an ability to regard everything on its merits without regard to customs, doctrine or drill. Nothing in this book can compare in importance with an understanding of this fact. An officer who understands the system is more use to his government and to his men than one who has spent years learning how to use the latest devices produced by modern technology.

Chapter 7

INSURGENCY Part II:
Direction, Units and Equipment

The last chapter was devoted to explaining a method of fighting insurgents which is based on collecting and developing information and which is relevant to any situation. The purpose of this chapter is first to show how a campaign should be unfolded in order that the most favourable circumstances can be produced for the use of this method of approach, and second to examine the type of units and the sorts of equipment most needed for taking advantage of it. The first part is relevant to the steps which the army should take to prepare itself, because it governs the education which officers should receive, especially at the higher level. The second part is relevant because it shows what type of unit and what sort of equipment is required by the army for fighting insurgents.

Direction of the Campaign

Grivas makes the point[1] that in any given type of terrain there is an optimum number of troops which can usefully be employed, and that if the number is increased beyond this point it helps the insurgents more than the government. He gives as his reason the advantages which the insurgents gain from the proliferation of targets, when there are too many of them. He might also have mentioned the extent to which it forces up the cost of the campaign because this was undoubtedly an important factor in securing victory for EOKA in Cyprus. Another important disadvantage of having too many men is that it tempts commanders into using them in vast hit or miss type operations which although they may temporarily disrupt the enemy's arrangements, and even cause casualties, fail to get at the roots of his organization. Worse still they break such continuity in collecting and developing information as the normally resident troops or police may have built up.

Undoubtedly Grivas was right to point out the dangers inherent in deploying too many troops, and it is almost certain that the

[1] GEORGE GRIVAS, *Guerilla Warfare* p.46.

Americans have suffered in Viet Nam from using force in an un-economical fashion in the same way as the British suffered in Cyprus. There is little doubt that one of the main problems facing government leaders in a counter-insurgency campaign is to work out how to use men and equipment as economically and effectively as possible and a geat deal depends on the skill with which this is done. Very often the problem of having too many resources does not arise, and the business of conducting the campaign is largely one of using such forces as are available in as economic a way as possible. But even if unlimited resources are available their economic use is none the less essential both because insurgents rely heavily on the cumulative costs of the war working in their favour, and because of the adverse effects of using force un-economically which have already been mentioned.

The ideal situation is undoubtedly one in which it is economi-cally possible to allocate just enough troops to each area to enable the committee or commander there to stabilize the situation and carry out such offensive operations as may be needed without drawing on reinforcements, and without having to send forces to a neighbouring area. A more economic method of deploying troops, and one which will still enable the development of infor-mation process to take place, is to take a chance in one area, whilst another one is pacified, and then move troops from the pacified area to the other one as soon as it is certain that the first area will not revert to the insurgents. The key to this business lies in the ability of the police and locally raised forces to hold the pacified area for the government when the soldiers move else-where. If a pacified area is allowed to slip back completely under insurgent control it will be more difficult to reclaim, because many of those sympathetic to the government will have shown their hand during the period of the government's ascendancy and will have been killed when the insurgents regained control. But this only happens if the area slips back completely. Very often, as in the scenario, the insurgents make some headway when the soldiers leave, but can still be handled when they return.

It should be noticed that whether or not it is necessary to abandon areas in pursuance of this plan, the system is essentially different from that used by the French in Algeria which was known as 'quadrillage'[1] and from the one used by the British in Malaya

[1] EDGAR O'BALLANCE, *The Algerian Insurrection*, p.64.

where small units carried out 'framework operations' until they could be reinforced by larger units sent to carry out a 'federal priority operation'.[1] Both these systems were based on the idea of keeping a few troops in an area all the time to provide some continuity and protection to the locals, but under both systems large-scale reinforcement was required before offensive operations could be carried out. Systems of this sort undoubtedly represent an improvement over the old fashioned idea of a destructive or punitive expedition into rebel held territory, because the framework part of the procedure recognizes the need for joint civil military action to organize the population and build up background information. The weakness lies in the second part of the concept, that is to say in the idea of boosting the framework force with larger units to carry out offensive operations, because the commanders of the reinforcing units outrank those of the framework units and as a result continuity of operational command is broken. The important feature of the idea advocated in this chapter is that the same forces carry out both framework and offensive operations without having to be reinforced, because the development of information process enables the operations to be carried out by far fewer men than are needed if the random approach is adopted. It can of course be argued that if troops are taken from one area and sent to another one, continuity of operational command is broken just as effectively as it would be under the other system, but this only happens once the area from which they are being taken is in a fit state to be controlled by the police and administration: it does not automatically happen immediately before a major operational push as is the case when reinforcements arrive to conduct offensive operations. Furthermore, if the optimum number of troops are available in the country as a whole it will not happen at all, and even this optimum is less than the number required to work the framework and reinforcement system.

If the strength of the two sides is evenly balanced, either because of government mistakes in the early part of the campaign or because the insurgents have strong support, the struggle becomes one of life and death. The government is then hard pressed to know which areas it should hold, and which it should abandon for the time being. The enemy will do all in his power to direct troops

[1] RICHARD CLUTTERBUCK, op. cit., p.112.

away from areas which seem important to him, by such means as fermenting urban disturbances in the hope that troops will be taken from country areas which can then be swallowed up. This forces the government to decide whether to suppress the urban disturbances violently but economically, thereby taking the risk of alienating world opinion, or whether to withdraw enough soldiers from other areas to regain control by gentler methods, thereby losing ground in the areas from which the troops are taken.

Successful direction of the campaign involves balancing all these risks and correctly gauging the moment for moving troops from one area to the next. It also involves knowing how to use such different types of support as that provided by artillery and air power in order to fill the gap in the deployment plan. For example, during the second half of 1954 in Kenya, General Erskine decided to use most of his soldiers in helping to raise and train Home Guard in the rural areas,[1] the situation being analogous to that depicted in the opening stages of the scenario. He did not want to leave any area completely untended which meant that no one District could find enough men to carry out the essential protectional and organizational task of building up the Home Guard, and at the same time find forces for offensive operations against enemy armed groups in the forest. General Erskine therefore put his bombers at the disposal of the District Emergency Committees so that they could harass the forest gangs with them until they were ready to resume offensive operations. Under certain conditions special forces are more effective or economic than regular troops in which case the overall plan should be geared towards utilizing them either in addition to, or instead of, regular troops. As in all forms of war it is important to know how to make the best use of the resources available, and in fighting insurgents the key to this knowledge lies in a correct understanding of the importance of building up and developing information. Effective direction of a campaign from a tactical point of view is likely to depend on the extent to which priority is given to this activity.

Units and Equipment

So far as the army is concerned, the task of fighting insurgents is largely one for the infantry although support from engineers

[1] ARTHUR CAMPBELL, op. cit., p.219.

and armoured cars is almost always required. Under certain circumstances a full scale of supporting arms, including artillery and fighter ground attack aircraft, may be needed from the start, as was the case in the Radfan Operations of 1964.[1] It is also likely to be required when a counter-insurgency campaign deteriorates to the extent where insurgents openly take the field. In fact, the problems involved in organizing, training and equipping the army to fight insurgents become less as the campaign approximates more closely to regular warfare, because the army is prepared for this in the normal course of events. In considering the question of the type of units and the sort of equipment which the army should be prepared to put into the field against insurgents it is therefore advisable to relate the problem to a situation as far removed from conventional warfare as possible.

An infantry battalion does not need to be reorganized or re-equipped to any great extent in order to fit it for fighting insurgents. Usually it can dispense with its anti-tank weapons, and it may require a different scale of vehicles and wireless sets but these are all matters of detail. Trackers are almost always of great value and there are no grounds for thinking that they can only be used if the campaign happens to take place in a primitive country where a proportion of the people track for their livelihood. Trackers from one country can perfectly well operate outside their own immediate environment provided that conditions are similar. During 1958 trackers from the Northern Frontier District of Kenya were used in the Sultanate of Muscat and Oman, and were so successful that a further contingent were asked for and sent in 1960 to deal with an outbreak of mining that occurred at that time. Wherever possible trackers should be attached to infantry battalions taking part in counter-insurgency operations in rural areas, and the soldiers should be taught to work with them. In addition to human trackers there is likely to be a requirement for tracker dogs in the infantry battalion and also for guard or patrol dogs in infantry and other units.

When looking at the need to prepare units of other arms to take part in counter-insurgency operations it is only necessary to say that although they may be used in this way from time to time, it is unlikely that they will be given special arms or equipment because of the cost involved. This is certainly the case in the British

[1] JULIAN PAGET, *Last Post : Aden 1964–67*, p.83.

Army at any rate, and to date most weapons designed for conventional war have been perfectly adequate for counter-insurgency operations because they have had to be reasonably mobile to meet the demands of the strategic reserve. Should any particular range of weapon, such as the medium gun, cease to be produced in a strategically mobile form, the question of buying a small number specially designed to meet the requirements of counter-insurgency operations would have to be considered, but it is unlikely that much money would be spent in this way unless there was a strong probability of the equipment concerned being wanted. At this juncture it is just worth touching on the problem of using armoured or artillery units in a dismounted role, to make good a shortage of infantry battalions. This has been done on several occasions in the past by the British Army, as for example in Cyprus in 1964 and in Northern Ireland in 1969 and the advantages and disadvantages are covered in some detail in Chapter 10. It is, however, necessary to point out that although units may be successfully used in this way for peace-keeping or for handling protest meetings, riots and other non-violent forms of disturbance, they can only be used in counter-insurgency operations after an extensive period of re-training and that even then they are unlikely to be very effective.

An aspect of counter-insurgency operations which needs constant attention is that which covers the use of aircraft. The advent of the helicopter in the 1960's was seen by many people as a major breakthrough in this field and helicopters undoubtedly increase the ability of a force to operate in a number of ways. But there are some important disadvantages to be considered in relation to their employment which also apply to a lesser extent to other aircraft. The first of these is that helicopters are relatively expensive in terms of money and the introduction of large quantities of them into a campaign pushes the cost up appreciably. Second, although they may enable a commander to use his fighting troops in an economical way, they themselves need a considerable number of highly specialized men to service and fly them. Third, the excessive use of helicopters by troops results in their living in a totally different medium from the enemy which is highly detrimental to their chances of catching him. This is particularly true of commanders who dash from place to place by air never giving themselves time to soak up the atmosphere which is guiding the actions of their opponents. Finally, although armed helicopters as developed by the Americans in Viet Nam can be highly effective against groups of insurgents operating away from populated areas, it is necessary to employ them with extreme caution in a place where there is a chance of causing damage and casualties to the civilian population. Accidents can play an appreciable part in embittering the population, especially if the impression is allowed to become established that the government is prepared to accept them for the sake of causing casualties to the enemy. It is therefore reasonable to view with some reserve the use of a weapon system which is likely to cause them in a campaign which is ultimately based on the need to regain and retain the allegiance of the people. It is probably true to say that the use of heavily armed helicopters in a populated area can only be justified if the campaign has deteriorated to the extent where it is virtually indistinguishable from conventional war, that tasks in which they can legitimately be used before that time, i.e. attacks away from centres of population, can often be carried out by aircraft designed to give support to ground forces in conventional war, that the British are unlikely to produce them specifically for counter-insurgency operations, and that unless used with great care and restraint they are likely to do harm as well as good.

The only proper way to assess the correct use of helicopters, and of other aircraft for that matter, is to see how they can be used to promote the overall campaign plan and the individual low level plans designed to collect and develop information. For example if the plan of operations in a district involves establishing a base camp in a forest or in a mountainous area, and if there is no intention of making a secret of the matter, then a large number of man hours can be saved by moving the soldiers into the camp by air and by supplying them from the air thereafter. Indeed many effective operations in the past could never have been undertaken in any other way. But if the district plan involves establishing a camp as a base for patrols and ambushes which relies for its success on being undetected, then re-supply from the air by brilliantly coloured parachutes would be likely to nullify the whole operation, and very many operations have been ruined in this way. Of course the situation is seldom so clear cut as this. Various measures can be taken to reduce the risk of air action giving away positions, and the remaining risks have to be balanced against the advantages gained. The important point in preparing an army to fight insurgents is that it is not enough to provide aircraft and teach people what they are capable of doing. It is equally important to teach them how to use them in the furtherance of the aim.

It has been shown how special units have a part to play in a counter-insurgency campaign and it is worth considering what steps are necessary for ensuring that the best use is made of them. In one way the situation is not unlike that relating to the use of aircraft, in that special units can only work effectively if they are tied into operational plans by people who understand what can be achieved by them, and who also understand the nature of counter-insurgency operations sufficiently well to realize when they can be used to advantage and when their use is likely to be detrimental to the conduct of the campaign as a whole. The major difference between the use of special units on the one hand and of aircraft or normal units on the other, is that, almost by definition, they can only be raised in the theatre of operations during the course of the campaign. The most that can be done in advance is to teach all officers to know how they should be used and to maintain particular individuals who can study the subject in detail and who can be held ready in every way to go to the scene and raise forces of this sort as quickly as possible.

The requirement for the army to have men trained to act as psychological operations advisers, and units ready to carry out the necessary specialist functions has already been made, and further reference to the way in which the requirement might be met will be found in the third part of this book. It is nonetheless worth raising the subject again at this point in order to stress its relevance to the counter-insurgency phase of operations. Not only must specialist advisers and units be provided, but the potentialities of this form of warfare should be taught to all officers who must understand how to exploit them to the full.

In trying to make the best use of available resources it is important that new developments in the field of weapons and equipment should be harnessed to the task of fighting insurgents as quickly as possible, but in this connection it is necessary to ensure that the use of the new device will promote the government's aim in the widest sense as well as being helpful in the immediate context in which it is used. A good example of a case where insufficient care was taken in this respect concerns the introduction of non-lethal gas into the Viet Nam war. Several attempts to use non-lethal gas in conjunction with orthodox assaults on Viet Cong positions were made early in 1965 but the practical results were virtually nil. But when Washington casually said in March 1965 that gas had been used in this way, the announcement resulted in considerable and adverse political repercussions around the world.[1] After 1965 the United States Army considerably improved its techniques for combining the use of non-lethal gas with fire from weapons to kill Viet Cong and by September 1968 it was using it as a matter of routine to drive the enemy into the open before artillery bombardments or bomber raids.[2] But despite the military effectiveness of this method, it is questionable whether the advantages gained outweighed the adverse public opinion reaction to its continued use. Certainly in the present state of world opinion it can not be imagined that Britain would use non-lethal gas in conjunction with conventional weapons as a method of killing, even if fewer casualties were sustained by the troops as a result. By contrast the use of temporarily incapacitating gases for riot control seems to have become accepted in most parts of the world.

The Viet Nam campaign has also seen the development of chemical measures designed to destroy vegetation, and it is possible that this technique will be used again in the future. Sometimes the process is carried out to restrict the amount of cover which enemy groups can get from woodland and scrub but there are serious disadvantages to the system. In the first place vegetation quickly grows up after a first spraying, but if sprayed again there is a 'heavy destruction of all woody plants'[3] which is unlikely to be popular in the eyes of the country's inhabitants or to be compatible with any long term government aim, except where the spraying is restricted to strips along the edge of roads. The second disadvantage is that although a defoliated area looks clear and open from the air, the amount of dead brushwood on the ground affords almost as much cover for the enemy and almost as serious an obstacle to movement as the original jungle. Sometimes defoliation is adopted for the purpose of destroying crops as opposed to cover, and in this case a different set of issues is raised.

[1] RICHARD TABOR, *The War of the Flea*, Paladin, 1970, p.146.
[2] RICHARD D. MCCARTHY, *The Ultimate Folly*, Victor Gollancz Ltd., 1970, p.49.
[3] Ibid., p.82.

A method of justifying this is to say that armies have traditionally destroyed crops and stores of food which might be of use to the enemy, and that destroying them with chemicals is no different from destroying them by fire. On the other hand it can be argued that chemical methods are so efficient that much more damage is done, that the long-term effects of spreading poison around the countryside are totally different from those resulting from fires, and that in any case the results of such widespread destruction will bring suffering to vast numbers of people who are not helping the enemy. Furthermore it will be said that there is little difference between this and ordinary poisoning. Irrespective of the logic of the case it seems unlikely that Britain will want to adopt such measures because of the adverse effect which they would have on world and domestic opinion in relation to the advantages gained.

An area in which technological developments may produce a very important advantage for those engaged in counter-insurgency concerns the development of night fighting equipment. Rural insurgents have always made maximum use of darkness to offset their weakness and attacks on the posts of government forces, ambushing, and food collection is normally carried out at night. The same can be said for the movement of messengers and commanders. Troops armed with portable radar, image intensifiers and FEBA[1] alarms will have a greatly increased capacity for countering enemy moves of this sort providing that the equipment is issued in sufficient quantities, and that the men are well trained in its use. So far as Britain is concerned priority will almost certainly be given to developing the equipment in relation to conventional war but fortunately there is no wide divergence in the conventional war requirement as compared to that for counter-insurgency. Even so it will be necessary to keep putting forward the needs of troops preparing for counter-insurgency operations because of the great importance of the equipment in this sphere.

Developments of new communications methods, although less spectacular, are none the less important from the counter-insurgency point of view; this is particularly true of devices designed to improve security because insurgents rely to a considerable extent on getting information from monitoring telephones and wireless links. In this sphere the requirements of counter-insurgency

[1] FEBA, Forward Edge of the Battle Area.

do not always coincide with those of conventional war, particularly with regard to the sort of wireless sets needed at the lower levels, because of the much longer ranges at which small groups of men have to operate, and because of problems posed by screening. Once again, the only thing that can be said is that in all phases of research and development the requirements of fighting insurgents should be given proper consideration.

In addition to the devices invented and designed for conventional war which are also useful for fighting insurgents, there must be many others invented and designed for industry or commerce which would also be useful if they could be identified and adapted to the purpose. No survey of the equipment which the army needs in order to make it ready to fight insurgents, would be complete without reference to these devices, and anything which would help in developing background information into contact information would be of particular interest. One way of approaching this subject would be to imagine a tactical problem and then see whether technology might be able to help in solving it. For example, under present arrangements, a member of the intelligence organization called to interrogate a prisoner in a forward area would be most unlikely to make any headway until he could get the man back to his office where he has the records of all the enemy groups in his area, together with details about their arms, contacts, and other points of interest. Even then if the prisoner comes from outside the area the interrogater would probably make little progress until information from the intelligence organization which covers the area in which the prisoner usually operates could be obtained. By this time a lot of the prisoner's immediate tactical value will have evaporated. From a purely technological point of view it would presumably be possible to solve this problem. All that would be necessary would be for a central computer to store all the information held in all the branches of the intelligence organization throughout the country, and for each member of the intelligence organization to be equipped with some form of wireless which would enable him to contact the computer from anywhere in his area. By this means the interrogater in the forward area could in theory get the information which he needs in order to break down the prisoner without delay. In practical as opposed to technological terms, the whole idea in the form suggested would almost certainly founder because of the cost, and because of the difficulty of teaching low level members of the intelligence organization how to work a computer by remote control in addition to all the other things they would have to learn. None the less some practical methods of helping members of the intelligence organization, based on the use of certain communications facilities together with some form of data storage mechanism, could probably be devised if the problem was being considered by a group of people who not only understood the needs of fighting insurgents, but who also knew what it was practicable to achieve in scientific terms. This calls for properly organized research and it is important that research should be carried out in the interests of fighting insurgents as well as in other fields.

Fighting Insurgency Summary

It is probably fair to say that throughout the armies of the world there is an ever increasing awareness of the importance of handling the population correctly in wars of counter-insurgency. The techniques of protection and organization are well known, and widely practised as a result of the lessons learnt in a number of campaigns such as Malaya and Viet Nam. On the other hand it is also fair to say that the armies of the world have been less successful in understanding the tactical problems concerned with destroying armed insurgents and the groups which support and direct them. As a result they do not always reap the rewards which they deserve from their understanding of the basic political issues involved. Chapter 6 attempted to explain a tactical method of approach designed to enable operations to develop effectively in accordance with the aim of the campaign. In this chapter mention has been made of a number of more or less independent factors which need attention if the army is to be ready to fight insurgents, but the main point relative to all the issues raised is that it is not enough merely to provide units, specialists, weapons and equipment as required, although all these things must be done. The really important thing is that people should understand how they can be employed within the framework suggested, in order to achieve the aim of the campaign. The best weapon employed other than in this context, is as likely to be a liability as an asset, and the most expensive equipment operating in a tactical vacuum, will probably help the enemy because it will put up the cost of the campaign without achieving any useful results.

Chapter 8

Peace-keeping

Although peace-keeping is a fundamentally different occupation to the countering of subversion, there is a surprising similarity in the outward forms of many of the techniques involved. On this account a certain amount of the preparation needed for fitting the army to carry out the latter task is also relevant to the former and this is the main reason for including the subject in this study. It is also important that those involved in countering subversion should realize that they are involved in this activity and not in peace-keeping, even when the outward forms are very much the same. It is not difficult to become confused in this respect, although it is unlikely that anyone genuinely involved in peace-keeping would consider himself to be taking part in a counter-subversion operation. The purpose of this chapter is therefore to establish the role of the army in a peace-keeping operation with a view to identifying the steps which should be taken to fit it for the task. In this connection the form of preparation once more divides itself on the one hand into the training and education which the officers of a peace-keeping force require, and on the other into the provision of the sort of units and equipment necessary for carrying out the task.

The term peace-keeping covers a wide variety of functions, not all of which include the use of military forces. For example both the League of Nations and the United Nations have on occasions exercised their influence to iron out problems between conflicting groups by providing impartial investigators on the assumption that it is only necessary for the true facts of the case to be ascertained by someone whom both sides trust, for a solution to present itself which both sides will feel bound to accept.[1] There are many examples of this device being used. One dating back to 1920 concerned the dispute between Finland and Sweden over the ownership of the Aaland Island at the entrance to the gulf of Bothnia which was settled in Finland's favour after investigation by a League of Nations Commission.[2] A more recent example can be

[1] ALAN JAMES, *The Politics of Peace-keeping*, Chatto and Windus, 1969, p.10.
[2] Ibid., p.16.

found in the opposition voiced by Indonesia and the Philippines to the proposal to include North Borneo and Sarawak into the new State of Malaysia on the grounds that such a move was contrary to the wishes of the people concerned. This was the subject of an investigation by a United Nations group in 1963 which decided that the inclusion of these territories into Malaysia could be regarded as being in accordance with the freely expressed views of their people.[1] Mediation is another device which does not involve the use of military forces. It has been defined as the attempt of an intermediary to draw disputing parties together and so obtain an agreed settlement.[2] Sometimes a mediator may operate in an area where a peace-keeping force is also operating as was the case in Cyprus during the period when first Mr Tuomioja and then Mr Galo Plaza tried to find a solution to the problems which were causing strife between the Greek and Turkish communities on the island.[3] It should be noted that neither of these men were part of the United Nations Force in Cyprus when they were acting as mediator but were directly responsible to the Secretary-General in New York. An example of a mediator operating in an area where no United Nations forces were deployed can be found in the appointment by the Secretary-General of the United Nations in 1962 of Mr Herbert de Ribbing to mediate between Great Britain and Saudi Arabia over the Buraimi Oasis dispute.[4]

Even when a peace-keeping force is involved, there is an immense range of activities which it may be called upon to undertake. So far as past situations are concerned these have included provision by the United Nations of five or six relatively small Truce Supervisory Organizations or Observer Groups such as that deployed after the 1965 war between India and Pakistan which numbered about 115 men.[5] A totally different sort of operation was undertaken in 1963 when the United Nations took over from Holland responsibility for West New Guinea and looked after it for some months before handing it over to Indonesia. This was done with the agreement of the two powers concerned in order to

[1] ALAN JAMES, *The Politics of Peace-keeping*, Chatto and Windus, 1969, p.32.
[2] Ibid., p.36.
[3] Ibid., p.73.
[4] Ibid., p.70.
[5] Ibid., p.119.

prevent conflict between them, and involved the deployment of a force of 1,500 men.[1] On two occasions the United Nations have put forces of around 6,000 men into the field, first in Egypt in 1956 following the Suez campaign[2] and then in Cyprus in 1964 to take over from the British, Greek and Turkish peace-keeping force which was set up to deal with inter-communal fighting on the island.[3] The largest United Nations peace-keeping operation took place in the Congo, where nearly 20,000 men were deployed in 1960.[4] The only other peace-keeping operation of comparable size in terms of manpower was that carried out in the Dominican Republic by the Organization of American States in 1965 when contingents from the United States, Honduras, Nicaragua and Costa Rica under the command of a Brazilian general intervened to save the country from anarchy.[5] In passing, it is worth pointing out that the Korean campaign, although fought under the aegis of the United Nations, was not an example of peace-keeping but was a straightforward exercise in collective security.

But although peace-keeping can take so many forms, and although it may involve the use of forces varying widely in strength and organization, it is absolutely different in its nature from all other sorts of operations. The essential difference is that a peace-keeping force acts on behalf of both parties to a dispute, at the invitation of them both, and therefore must as far as possible carry out its task without having recourse to warlike action against either of them.[6] It follows that the body sponsoring the force can not be responsible for the government of either of the parties to the dispute, because if it were, it would not be in a position to act on behalf of the other one, nor would it be invited to do so: if the body is responsible for the government of one of the sides the operation becomes one of ordinary war, and if it is responsible for the government of both sides it becomes one of subversion, insurgency or civil war. The fundamental characteristics of peace-keeping operations arise very largely out of this factor.

[1] ALAN JAMES, *The Politics of Peace-keeping*, Chatto and Windus, 1969, p.160.
[2] Ibid., p.99.
[3] Ibid., p.324.
[4] Ibid., p.355.
[5] Ibid., pp.239–240.
[6] See above, pp.3–4.

The other main factor, and one which is closely related to the first, is that the terms of reference which govern the way in which the force operates are often far less precise than is desirable from a military point of view. There are two reasons for this. First, a sponsoring body such as the United Nations or the Organization of American States consists of many separate countries each of which may have its own idea as to exactly what the peace-keeping force should do according to how it views the rights and wrongs of the dispute. Second, the mandate has to be acceptable to both parties and therefore has to be framed in such a way as to give no advantage to either side. For both these reasons it is bound to be imprecise and full of holes so that all sorts of different interpretations can be placed on it by the two parties involved and all those contributing to the force.

A good example of the complicated considerations which can develop around the establishment of a peace-keeping force is afforded by the events which took place in Cyprus during the winter of 1963–64. In this case rivalry between Greek and Turkish Cypriots had resulted in an outbreak of inter-communal fighting. Although the Greek Cypriots realized that they could easily beat the Turkish Cypriots, the also realized that they were in grave danger of being invaded by mainland Turkey. They did not want to concede to the demands of the Turkish Cypriots nor did they want to be invaded by Turkey so, as a last resort, they agreed to the establishing of a British peace-keeping force. In theory it might have paid the Turkish Cypriots to have refused to accept a peace-keeping force and to have let the trouble develop to the extent where Turkey would have been obliged to come to their rescue. But to have done this would have resulted in the death of a large number of their people at the hands of the Greek Cypriots before the rescue could have become effective which made the idea of a peace-keeping force attractive to the more moderate members of their community. More important, it was clear that a Turkish invasion would inevitably lead to a war between Greece and Turkey which although it would probably result in a win for Turkey if it was allowed to run its course, would greatly weaken the NATO position in the Eastern Mediterranean. Turkey was therefore subjected to pressure from America and other NATO powers to agree to a peace-keeping force being established as an alternative to invasion, and she in turn obliged the Turkish

Cypriot leadership to accept it. Finally Greece accepted the idea because she had no wish to go to war with Turkey and in any case she could hardly have opposed it when each of the other three parties to the dispute had agreed. Thus, as the result of complicated negotiations and for a wide variety of reasons, all those involved decided that it would be advantageous for a peace-keeping force to be established, but this did not mean that the underlying problem had been solved. On the contrary all four parties had every intention of continuing to pursue their interests but they were prepared to do so at a lower level of intensity and they therefore engaged the peace-keeping force to damp things down when they looked like getting out of control.

The force moved into position swiftly and before proper terms of reference could be worked out. The rules were therefore evolved over a period and were initially adequate for enabling the force to control the situation. But it was soon apparent that Britain would not be able to handle the matter indefinitely because of the opposition which was building up as the force thwarted the machinations of first one side and then the other in the course of its duties. Furthermore Britain had no intention of shouldering the burden indefinitely and the end was obviously not in sight. Various alternatives were considered but it was soon clear that only a United Nations force would have any chance of being accepted, and that there would be considerable difficulties involved in working out terms of reference which could be agreed by the parties concerned and by a satisfactory majority of the members of the Security Council including all those in a position to exercise the veto. It is hardly surprising that the mandate which was finally issued to the force was worded in a very general manner and in one which was capable of being interpreted in several ways. The most positive clause of the Security Council Resolution establishing the force, was one which recommended that the function of the force should be to use its best efforts to prevent a recurrence of fighting and, as necessary, to contribute to the maintenance and restoration of law and order and a return to normal conditions.[1]

Although the position in Cyprus during the first few months of 1964 was extremely complicated, it was not untypical of the sort of

---

[1] Security Council Resolution S/5575 of 4 March, 1964.

situation which often confronts those involved in peace-keeping. The United Nations Force in the Yemen was confronted by four parties, i.e., Yemeni Republicans and Yemeni Royalists, backed by Egyptians and Saudi-Arabians respectively, each of whom had their own backers in the United Nations.[1] In the Congo the United Nations force ostensibly moved in to help facilitate the withdrawal of Belgian troops who had returned after the country had gained its independence because of the anarchy which spread in the wake of a mutiny of Congolese troops.[2] But in practice a more pressing reason seems to have been the hope that United Nations involvement would prevent other powers coming to the 'assistance' of the Congo and thereby importing international strife into an area already sufficiently inflamed.[3] Later on the United Nations, having tried to avoid intervening in the internal disputes which arose amongst the Congolese themselves, found itself obliged to do so on behalf of the central government against the secessionist province of Katanga because a number of influential UN member states demanded it, some threatening to withdraw contingents serving in the force if this particular task was not carried out. One country even considered withdrawing its contingent from the United Nations force and allying it with the Congolese government in an attack on Katanga.[4]

In order to understand the pressures which are likely to confront officers concerned in the direction of peace-keeping operations it is necessary to realize the overriding influence of the political intricacies involved, and in preparing officers to take part the first essential is to get this point accepted. Although the political background may be more significant at the level of the force and contingent commanders than it is to those below them, it is none the less highly relevant to company and platoon commanders, not only because it often governs their plans, decisions and negotiating activities, but also because only by understanding it can they explain otherwise incomprehensible actions and orders to their men. All too often action or inaction is necessary from a political point of view in a peace-keeping operation which runs contrary to the sort of action which soldiers would take in war:

[1] ALAN JAMES, op. cit., pp.108–110.
[2] Ibid., p.355.
[3] Ibid., p.357.
[4] Ibid., p.415.

L

it may even run counter to that most important military attribute, an offensive spirit, which is carefully inculcated into all soldiers in order to fit them to fight battles rather than to umpire them. This point which has not hitherto received much attention in written works on peace-keeping operations is of great importance. In effect it means that for political reasons the demands placed on soldiers in peace-keeping operations can be extremely exacting and only capable of being met by well prepared and well disciplined troops. A by-product of this state of affairs is that soldiers who have been employed on peace-keeping tasks involving curtailment of their offensive spirit should not be employed in warlike operations until they have been extensively re-trained, a point also made in relation to soldiers employed in the control of non-violent subversion.[1]

An example of the pressures which troops in a peace-keeping force may be obliged to face for political reasons can be found in the events connected with the attack by Greek Cypriot forces on Turkish Cypriot positions in and around two Turkish villages in November 1967. In this case the original cause of the attack was the prolonged and unjustified refusal of the Turkish Cypriots in the area to permit Greek civil police to use a certain stretch of road in the execution of their duty. After considerable negotiation the Greek Cypriots decided to use military force to open the road. The United Nations at the highest diplomatic level had for weeks been using its influence to persuade the Turks to open the road and it also used its influence to prevent the Greeks from opening it by force, when they announced their intention of doing so. But although this exercise of influence was extensive at every level from the Secretary-General in New York to the company commander on the ground, the United Nations force was not prepared to use military measures either against the Turks to force them into opening the road or against the Greeks to stop them from opening it.[2] Either course would have been contrary to the idea of a peace-keeping force exercising its function on behalf of both sides to the dispute without recourse to warlike measures against either of them. In fact there were various courses of action open to the United Nations units at the time, but it is not necessary to discuss

[1] See p.90 above.
[2] M. HARBOTTLE, *The Impartial Soldier*, Oxford University Press, 1970, pp.158–159.

them here, because this example is concerned with pressures on
the troops. Suffice it to say that the orders given to them in the
event of an outbreak of fighting were to stand fast, observe, report
but not to intervene. The soldiers of the United Nations force were
scattered in groups of six or eight in a number of observation posts
around the area when the Greeks moved in to open the road.
Surprisingly the Turks made no hostile move when the Greeks
drove along the forbidden stretch on three occasions but the
Greeks had prepared a massive operation and were not going to
stop using the road until their rights had been thoroughly established or until the Turks had been provoked into firing on them.
On the fourth occasion the battle started and the Greeks launched
a massive attack against the whole Turkish position which
included an assault on the two villages.

During the course of the ensuing battle a number of small
United Nations posts were shot over, mortared, and then occupied
by Greek forces ranging from a platoon to a company in strength,
because they happened to be in the line of advance. In order to
have good observation the posts were sited in exactly the positions
which the Greeks wished to occupy in the course of their attack.
In two places Greek forces looted personal effects of United
Nations soldiers which they found in tents sited in sheltered
ground a hundred yards or so from the observation posts and in
another they tried to interfere with a wireless set to prevent
reports being sent back to company headquarters. In order to
carry out the task of maintaining observation and restoring the
situation the United Nations soldiers had no alternative but to
permit the looting of their personal effects although such treatment represented considerable indignity to the soldiers concerned, coming as it did, not from an enemy but from people whose
best interests the United Nations were supposed to be safeguard-
ing. In the case of the attempt to damage the wireless set the
United Nations soldiers were obliged to take action in order to be
capable of continuing their reporting and in this case the wireless
operator attacked his assailants with a bayonet. Although one
Greek officer was slightly hurt, fire was not exchanged. The whole
battle was altogether distasteful to the United Nations soldiers
concerned, not only because of the hostility displayed towards
them personally but also because it ran counter to their military
instincts to sit and watch a superior force attack a grossly inferior

one and then clear two villages of their civilian inhabitants. But the task of observing and reporting was carried out to the letter so that the Secretary-General in New York was in possession of the exact facts throughout the battle within a few minutes of events taking place. As a result top level diplomatic pressure could be brought to bear on the heads of all the governments concerned. The Chief of Staff of the United Nations force gives it as his opinion that the presence of the United Nations soldiers probably did more than anything else to limit casualties and to influence the Greek Government to order the calling off of the attack and the withdrawl of the forces from the area.[1]

This was of course an extreme situation and troops are seldom obliged to undergo such an ordeal for political reasons. Indeed it is not always necessary to remain passive as will be explained later in this chapter: it just happened to be what was required in this particular case. None the less political considerations do govern the efforts of those involved in peace-keeping to a much greater extent than they govern the efforts of those involved in most forms of warfare and the lesson has to be taught and learned. In some ways this problem is more acute in a well trained and operationally experienced army than it is in a conscript force. It would seem that the British have a great deal to learn judging by the difficulty which they apparently found in adjusting themselves to peace-keeping in Cyprus compared to the Canadians, Irish, and Swedes, according to the United Nations Chief of Staff there.[2] General von Horn makes the same point rather less bluntly when discussing the attitude of the British General Alexander who was commanding Ghanaian troops operating in the Congo in 1960. He states:

' . . . his [General Alexander's] proposals made extremely sound sense and would have been unhesitatingly adopted in any normal army. Unfortunately we were not an army, we were a United Nations Force in which logic, military principles – even common sense – took second place to political factors . . . I remember wondering whether one of the international soldier's hardest lessons is not to grasp the difficulties and pitfalls which encompass UN service'.[3]

[1] M. HARBOTTLE, *The Impartial Soldier*, Oxford University Press, 1970, p.157.
[2] Ibid., p.46.
[3] CARL VON HORN, *Soldiering for Peace*, Cassell, 1966, pp.146–147.

It is not necessary to accept the estimate of British adaptability in Cyprus or to agree with General von Horn's assessment of the particular views put forward by General Alexander (which might well have been right even in the peace-keeping context) to see that the problem of understanding the influence of political factors is absolutely fundamental to any officer concerned with peace-keeping.

Once the essentially political nature of peace-keeping has been accepted the next thing to consider is the way in which the political and military direction of events can be tied together. In this connection it is necessary to stress once more that the purpose of the peace-keeping force is usually to prevent fighting so as to gain time for the parties to the dispute to sort out their problems: the actual business of sorting out the problem is not itself part of the aim of the peace-keeping force although it may well be the aim of a mediator appointed by the body which is sponsoring the force. But even the business of winning time by preventing outbreaks of fighting is largely a political matter so that the commander of the force is likely to have a political adviser at the least and may even find himself working under the direction of a civilian representative of the sponsoring body. This at least is what happened in two out of the three places where large United Nations Forces were deployed, i.e. in the Congo and in Cyprus. In both of these places the Secretary-General of the United Nations appointed a civilian to be his Special Representative and this procedure is one that should be understood by all involved in peace-keeping. In practice the exact sphere of the military and of the political leader of a peace-keeping force is difficult to define and may depend largely on the personalities of the men concerned. It is reasonable to assume that high level negotiation will usually be carried out by the political representative of the sponsoring body if one exists with the force commander in attendance to comment on the feasibility from a military point of view of any of the ideas discussed.

But although at the level of force headquarters it may be difficult to separate the functions of military and political leadership the same problem is unlikely to occur at subordinate levels where there will not be any political representative at all. Here an ordinary army officer is likely to find himself responsible for all aspects of the business and it would undoubtedly be helpful for him to arrive with a basic understanding of what is expected and a background knowledge of the different functions involved. It is not possible in a study of this length to include a description of all the various techniques which have been tried in past peace-keeping operations any more than it is possible to do so for the techniques of counter-insurgency, but it is necessary to indicate in very broad outline the problems which are likely to occur so that attention may be directed towards the main areas relevant to preparing officers for the task. In general terms the work of a commander in a peace-keeping operation falls under two main headings and it may be helpful to look at each of these in turn. The first concerns the methods which he can use to prevent outbreaks of violence and to extinguish conflict if it does break out. The second concerns the problem of running the force under the sort of conditions which are likely to arise in a peace-keeping oper-ation. It might seem unnecessary even to mention the second problem because it could be argued that it is such a routine func-tion of command, but there are some special aspects relevant to peace-keeping which warrant attention.

As a rule the most effective method of preventing trouble, or of damping it down should it break out, is by negotiation. Providing a commander knows what is happening in time, he can often fore-stall or terminate a disturbance by getting in touch with the rival leaders and persuading them to modify their plans or activities. There are usually plenty of ways in which a commander can bring pressure to bear such as enlisting support from a person on the same side as the leader concerned but who operates at a higher level, or who is more influential for some other reason. Providing that the leaders at the very top do not want trouble to break out it will be averted altogether or stopped quickly. The most difficult outbreak to prevent is the spontaneous one at ground level but for obvious reasons it is also the easiest to control. If negotiations are ineffective by themselves, the next way in which a commander can influence events is by using the troops at his disposal, and this can often be done without the use of force itself. For example, if a commander discovers that one of the sides intends to occupy a particular piece of ground, and if he realizes that the other side is bound to oppose such action, he could occupy the ground with his own men. If it comes to using force as opposed merely to deploying soldiers, the whole business becomes very much more complicated, and will hinge on the agreement which was reached with the parties concerned when the force was originally set up. Very often peace-keeping forces are restricted to using force only in self-defence, or as a means of enforcing a specific agreement entered into by both sides, but this is not necessarily the case, and it is both logical and conceivable that the sponsoring body would only agree to the force being established providing that it was authorized to use whatever means were necessary for the achievement of the aim.

In theory it might seem desirable that a peace-keeping force should have as much freedom of action as possible in this respect, but in practice there are advantages in avoiding the use of violence when operating in a peace-keeping role. The reason for this is that no matter how well justified the use of force might seem to a dispassionate observer, it will not appear in the same light to the people who are being shot at—particularly if someone is killed. Once the men of a peace-keeping force have actually shed blood, the hostility which they are bound to encounter in any case, will become much more intense. Furthermore mistakes are often made in a tense situation and they can not be reversed if somebody dies as a result. A more important reason for avoiding the use of force, other than that required for self-defence, is that it runs counter to the conception of the peace-keepers acting on behalf of both sides, without recourse to warlike action against either side. There is a body of opinion which considers that a peace-keeping force should be strong enough in terms of numbers and equipment to ensure compliance with its orders in the pursuit of its mandate by force if necessary, but, apart from being impractical in relation to the size and therefore the cost of the force itself, such an interpretation of peace-keeping would immediately founder because the disputants would not invite such an army on to their soil,[1] and no mandate acceptable to all parties concerned could be drawn up which would be sufficiently precise to enable the commander of the peace-keeping force to use his men in this way. All that would be achieved in the unlikely event of such a situation coming about would be the introduction of a third party to the dog-fight.

But this is not to imply that those involved in peace-keeping should merely negotiate and then sit back and hope for the best. However energetically the negotiating has been carried out there

[1] M. HARBOTTLE, op. cit., p.47.

is always the possibility that nothing will come of it and that fighting will break out, in which case the peace-keeping force may suffer a blow to its prestige, and its credibility with all concerned may be weakened (although this is not necessarily the case if it is clear from the circumstances that no alternative exists and that both sides are determined to have a fight, as was the case in the example of events in Cyprus in 1967 quoted earlier). The real test of a commander's skill lies in using whatever powers are allowed him, in accordance with the rules which have been agreed. Even when he is limited to using force in self-defence there is a lot which can be done, and a certain amount of latitude can be expected in interpreting the phrases. For example a detachment of a peace-keeping force can occupy a bit of ground if it is thought necessary to do so, in order to prevent one or other of the sides from fighting over it. In this case if one of the parties advances as if to occupy the position, it can in the last resort be fired on in self-defence, providing that there are enough members of the peace-keeping force present to defend their position if the attacking party persists in its advance. In different circumstances the peace-keeping force might interpose itself between the two sides in such a way as to oblige them to cease their operations for fear of firing on it. If they do shoot, the fire can be returned in self-defence. Incidentally this is an extremely delicate operation because of the risk that the peace-keeping force will be pinned down in such a way that the danger does not justify a return of fire, but which prevents the men from moving and thus influencing events elsewhere. Very often the best use which a commander can make of his troops is to have them positioned in such a way that they can see what is going on and pass the information back. As a rule neither side will want to start a fight unless they can pin the blame for it on to their opponents, so the presence of impartial observers will provide a strong incentive for calling off the project. If in spite of this presence an attack is launched, the peace-keeping force will be in a position to provide excellent material for those trying to negotiate a cease-fire.

The ability of a commander to negotiate depends to a considerable extent on the information at his disposal and it is not only required with regard to the subject matter of the negotiations: it is also necessary in order to know who is the best man to talk to in any given situation. All too often in a tricky situation, particularly at the lower levels, the best results come from talking to someone other than the official spokesman put forward and a great deal depends on discovering whose influence counts for most in any particular direction. Similarly information about the intended actions of both sides is of great importance to a commander who is trying to deploy his troops in order to prevent an outbreak of violence. It would therefore seem reasonable to assume that a peace-keeping force should have a first class intelligence service. Unfortunately even this elementary deduction looks different when viewed from the angle of peace-keeping politics, because it is argued that collecting information about people who do not wish to provide it is a hostile act and that as the business of collecting it by covert means involves deceit, it destroys the trust which both parties should have in the peace-keeping force.[1] On the other hand the very fact that both the contesting parties will be trying to pursue their divergent aims means that they will try and exploit the presence of the peace-keeping force and trick it from time to time. They are therefore bound to try and conceal their plans from it on occasions, and it would be naïve for the peace-keeping force to rely on the parties to the dispute providing all the information which it requires in order to thwart their moves.

This poses a serious dilemma in that commanders cannot operate effectively without information but at the same time the collecting of it by covert means at any rate, is regarded as an hostile act by the contesting parties and may even be forbidden by the body sponsoring the peace-keeping for political reasons. Certainly the United Nations force in Cyprus was careful not to develop intelligence sources or cells other than its own eyes and ears and it relied entirely on overt information picked up honestly from what it saw or was told.[2] General von Horn also makes the point that intelligence was a 'dirty word' in the Congo although he openly admits that his force gleaned valuable information from monitoring wireless sets used by the Congolese army.[3]

It would seem that the answer to the dilemma must depend on the circumstances. If intelligence activities are permitted a high priority should be given to providing as good an intelligence

[1] M. HARBOTTLE, op. cit., p.29.
[2] M. HARBOTTLE, op. cit., p.28.
[3] CARL VON HORN, op. cit., p.204.

organization as possible which can exploit to the full information collected by overt means, but which also uses men highly skilled in the development of covert sources who can be directly attached to the force as intelligence officers or who can be inserted in the theatre of operations under some other suitable cover. But if, as is more likely, intelligence activities are not permitted, recourse must be had to information gained solely by overt means although there is no reason why this should not be systematically collected and efficiently collated. It is amazing how much useful information can be obtained in a perfectly open way providing these two functions are properly carried out. Trade and telephone directories are full of interesting facts and the layout of field cables or the movement of particular vehicles, can reveal significant organizational details. Journalists are another prolific source of overt information and can usually be persuaded to pass it on in return for help in getting to the scene of trouble quickly or for the occasional tip off that something is in the wind. Embassy staffs often know interesting details which they may be prepared to pass on. There are many similar methods of collecting overt information in a way which cannot possibly be regarded as hostile acts if carried out with discretion, but by far the most important source of it is the discussions which all commanders automatically have with the leaders of the two sides. In this connection the important thing is that the information should all be recorded and presented in such a way that discrepancies or corroborative evidence can be seen and evaulated for what it is worth.

In mechanical terms, a commander who is trying to achieve his aim by negotiation plus the use of his men, will rely to a great extent on communications. Good communications are an essential part of knowing what is going on and being in a position to influence events. It is difficult to solve the problem of providing adequate communications for a peace-keeping force before it is deployed, because so much depends on the circumstances of the operation. In one theatre the difficulty may lie in the great distances involved, whereas in another, or possibly in a different part of the same one, the difficulty is screening in urban areas. Furthermore there may be a good civilian telephone system which can be used within the limits imposed by security, or there may not. The only thing that can be foreseen with reasonable certainty is that normal military communications are unlikely to be adequate because they are designed to cater for a totally different sort of troop deployment. Any nation that contemplates providing contingents for peace-keeping forces should make some provision for handling this problem speedily, if only by ear-marking a realistic sum of ready cash so that suitable equipment can be bought once the requirement is known.

The problems involved in the day to day running of the force can conveniently be considered under the headings of morale, discipline and administration. From the point of view of morale the main difficulty arises from the fact that from the soldier's point of view the task is usually boring and exacting, requiring long period of vigilance. Successful peace-keeping provides little satisfaction for the soldiers of the force because it merely means that nothing happens which in turn means more boredom. When compared with the excitements of a successful encounter in a counter-insurgency campaign this leaves a lot to be desired, and it is extremely difficult to keep soldiers happy when achievement is so intangible. In many respects operating in the midst of violence and confusion is even less satisfactory because the soldiers may be obliged to stand in the middle of the fight whilst their activities are confined to making reports. On these occasions they may be subjected to considerable danger and humiliation as described in the example given in the earlier part of this chapter without having the satisfaction of fighting back. The only way in which a commander will be able to maintain good morale under such circumstances is to prepare the men as well as he can in advance so that they are not taken unawares by the demands put on them and then to keep in close touch with them explaining the purpose of each phase of negotiation and operation so that the men can understand how their contribution is helping to achieve the aim of the force.

There are two disciplinary problems which are particularly relevant to peace-keeping operations. The first one is that disorderly conduct in centres of population may provoke just the sort of spontaneous violence which it is the aim of the peace-keeping force to prevent, and at the least is likely to undermine confidence in the force itself. Such behaviour therefore has to be treated in a more serious light than would normally be the case. The second problem is far more difficult and concerns the temptations put in the way of individual soldiers of the force to help one or other of the sides by providing arms or information or by carrying messages or even people in peace-keeping transport under the immunity normally accorded to the force by the parties to the dispute. Such actions often get brought to light by the other side, frequently in such a way as to cause acute embarrassment in the peace-keeping force which may be subjected to restrictions as a result which in turn may mitigate against the force's ability to achieve its aim. Unfortunately those intent on bribing individuals of the peace-keeping force often have such large resources as to make the temptation very great indeed. It is important that commanders should be aware of the threat and devise whatever means they can to control it, such as pointing out that the prospects of being caught are high and the punishments severe. General von Horn is particularly conscious of the dangers arising out of corruption in a peace-keeping force and devotes a whole chapter of his book to the subject.[1]

Administrative problems are difficult to foresee and may not arise to any great extent especially if the peace-keeping force is a national one or one that is administratively supported by one nation as is the case in Cyprus where the British administer the United Nations force. On the other hand if the peace-keeping force is multi-national administrative conditions in the early stages may be chaotic. That at least is the lesson which comes out very clearly from the Congo and Yemen operations. For this reason a country sending a contingent to join an international peace-keeping force should be in a position to ensure that it will be properly looked after, either by convincing itself that the logistic arrangements of the sponsoring body will work or by making its own contingent logistically self-supporting.

There is little to be said about special organization or equipment required by troops taking part in peace-keeping operations. Because the procedures used are so different from those used in other operations it is perfectly possible to use armoured or artillery units in a dismounted role providing that they are properly briefed. The task will seem no stranger to them than it does to the infantryman. On the other hand such a course is not likely to be economic because of the higher overheads per man on the ground as compared with an infantry battalion. It may be of course that the body sponsoring the force does not want infantry

[1] CARL VON HORN, op. cit., pp.98–114.

but would prefer armoured cars, engineers or logistic units in which case these units would have to be tailored to the particular requirement as far as possible. In terms of equipment it is only necessary to say that support weapons are unlikely to be needed although they might be taken as a precaution, but that all devices designed to improve surveillance particularly at night are likely to be useful.

It is probably fair to summarize the steps which a country should take in order to be ready to send a contingent to a peace-keeping force in the following way. First, it must ensure that all officers are thoroughly educated in the background to peace-keeping so that they understand that the whole approach is totally different to any other sort of operation. Second, it should ensure that techniques which have been used in past peace-keeping oper-ations are analysed and studied in the appropriate officer training establishments, and in this connection particular emphasis needs to be placed on teaching how to collect and make the best possible use of overt information on the grounds that there may be no other intelligence on which to base planning or negotiation. Third, steps must be taken to ensure that suitable communications and logistic support is available.

Finally there is little doubt that only well prepared and highly disciplined troops will operate effectively in a peace-keeping role. The stress placed on officers and men alike is considerably greater than is popularly supposed to be the case. In fact it is probably true to say that the demands of peace-keeping constitute one of the greatest tests which a commander can experience and certainly one of the least agreeable.

PART THREE

PREPARATION REQUIRED

PART THREE

Chapter 9

Education and Training

The first part of this study was designed to show that the fighting of subversion and insurgency and the carrying out of peace-keeping operations are standard tasks which any army should be prepared to undertake, and that so far as the British Army is concerned they are tasks which are more likely to arise than the fighting of conventional wars. Part One was also designed to give a very brief introduction into the nature of subversion and insurgency with particular emphasis on the fact that military measures only represent one aspect of the problem. Part Two was written to direct attention at what is required of the army in the various phases of fighting subversion and insurgency and in peace-keeping operations. The purpose of Part Three is to draw together conclusions from the first two parts which are relevant to the steps which the army should take to prepare itself for the task, and to highlight those areas in which a new approach might be of value. The substance of Part Three is contained in two chapters, the first of which concerns recommendations regarding the way in which education and training should be tackled, and the second of which contains some suggestions about the provision of military units and specialists, and the development of the weapons and equipment required. A final chapter, summarizing the main conclusions of the study as a whole, is also included in Part Three.

In terms of subversion and insurgency the educational and training requirement for the army as identified in this book can be seen to fall into four separately identifiable parts. In the first place there is the genuinely educational function of attuning men's minds to cope with the environment of this sort of war. This is something which concerns all officers and through them the soldiers themselves. It involves explaining the fundamental nature of subversion and insurgency with particular reference to the way in which force can be employed to achieve political ends, and the way in which political considerations affect the use of force. Study of the fundamental nature of conflict has always been recognized as being an important step towards the understanding of conventional war, and it is no less relevant when applied to subversion and insurgency. There is now a vast amount of written work on the subject which deserves to be analysed and the result of such an analysis would be of value in many different ways. For example it would help military history students to study campaigns of insurgency and it would also help those interested in the principles of war to see how these principles are as applicable to subversion and insurgency as they are to other forms of conflict.[1]

The second aspect of training and education which is relevant to subversion and insurgency, concerns the way in which officers are taught how to put a campaign together using a combination of civil and military measures to achieve a single government aim. Training of this sort is necessary in order to fit officers to take part, together with representatives of the police and the civil government, in running campaigns at their respective levels. It involves teaching them the operational value of non-military ways of harming the enemy such as resettlement schemes and food control. If the potential for good of using such means in pursuit of the aim is not understood by the military representative in an area, they may never get tried, and likewise if the military member is not in a position to point out the drawbacks attendant on their use, they may be employed at the wrong time merely because a policeman or government official has heard of the method concerned being put to good use elsewhere without ever having understood the full implications of using it.

The third aspect involves teaching officers how to direct the activities of their own soldiers including of course any policemen or locally raised forces as may be put under their command. It concerns first and foremost the fostering of the idea that commanders are the people responsible for collecting together and developing information to the extent necessary for achieving their tactical aims. This method of approach as described in Chapter 6 is at present scarcely understood, although it is vaguely perceived by many of those who have taken part in counter-insurgency operations. Unfortunately there is very little teaching material

---

[1] The principles of war are particularly easy to demonstrate in relation to counter-insurgency operations as can be seen by relating them to the situations depicted in Chapters 6 and 7, where the relevance of economy of force, the problem of striking a balance between the needs of offensive action and security, and the value of co-operation between soldiers, policemen, and government officials is particularly well brought out.

available on this aspect of the subject. As mentioned, there is an ever increasing volume of literature available about the nature of insurgency, and military training pamphlets give adequate coverage of the techniques which have been found useful in the past. There is none the less a need to bridge the gap between by explaining the right sort of tactical framework in which to use the techniques to the best advantage.

The fourth aspect of education and training concerns the methods used to teach all ranks the actual techniques themselves, that is to say the best methods for carrying out patrols and ambushes, organizing convoys, preparing defensive bases, attacking enemy camps and all the many similar military activities which make up the business of fighting subversion and insurgency. This is the function of military training in its most conventional sense, and there is plenty of teaching material available. The problems involved in carrying out this aspect of training successfully concern first selecting the right lessons to stress, bearing in mind the amount of training time available, and second, setting the training in such a way that it makes sense in the context of the proper handling of information. In other words, although the procedure for laying an ambush could perfectly well be taught entirely out of context, a more satisfactory result is likely to be obtained if it is done within a realistic and instructive framework.

In addition to these four aspects of the training required to prepare men to take part in combating subversion and insurgency, two further ones can be identified which concern peace-keeping. The first of these relates to the teaching required in order to give officers, and through them the soldiers, an understanding of the fundamental nature of the task. The second is concerned with teaching techniques which have been used in past peace-keeping operations with particular emphasis on teaching commanders how to collect and make the best possible use of overt information. Although there is less written material available on this subject than there is on insurgency, there is still a respectable amount, and it is being added to all the time. In June 1967 a bibliography of peace-keeping[1] was published which gave descriptive reviews of 75 books, pamphlets, and reports and of

[1] ALBERT LEGAULT, *Peace-Keeping Operations: Bibliography*, IPKO Publications, Paris, 1967.

167

approximately 90 important United Nations documents. Since that time a great deal more has been published so that it cannot be said that there would be any difficulty in finding as much material to study as there is time in which to study it. In a paper published in May 1967 the principal military adviser to the United Nations Secretary-General particularly stresses the need for countries which feel that they may one day contribute to a United Nations Force to include the study of peace-keeping activities in their officers' training.[1]

It is now necessary to examine briefly the ways in which these subjects should be taught so that the best value can be obtained from the time and money expended. In 1960 the British Army, realizing that regimental officers were constantly being committed to counter-insurgency operations without proper training, decided to run a special course designed to fit them for the task of commanding companies in this sort of war. At the time it was realized that the key to fighting insurgents lay in the proper handling of information and that a lot of the course would have to be concerned with explaining how background information is obtained so that it can be developed. For this reason it was decided that the course should be run at the School of Military Intelligence and not at an establishment designed for the teaching of tactics such as the School of Infantry. Although the course was a good one and well suited to the requirement, and although the reasons for holding it at the School of Military Intelligence were perfectly sensible ones, it foundered within a few years because of lack of support. Commanding officers, not having been brought up to understand the true nature of subversion, were unable to realize that the course was essentially tactical, and thought that they were being obliged to forego the use of their badly needed company commanders for ten weeks merely so that they could learn about intelligence. Had the course been run at the School of Infantry this misunderstanding might have been avoided but the course would probably have been less valuable. Be that as it may, the first serious attempt to come to grips with the problem ended in failure.

But in fact, although the idea was worth trying in the circumstances prevailing at the time, a course lasting for a few weeks

[1] I. J. RIKHYE, *United Nations Peace-Keeping Operations: Higher Conduct,* IPKO Publications, 1967, p.16.

and aimed at a particular level of officer could never have been anything more than a stop-gap measure because the real requirement is to teach the subject progressively to all officers as they pursue their careers. Fighting subversion or insurgency is no more of a special subject than is the fighting of conventional war. It is all part of the same subject, i.e. fighting, and the only rational way of approaching the problem is to teach it as a perfectly normal phase of war. Thus, aspects relevant to cadets must be taught at cadet schools such as Sandhurst, and aspects relevant to staff officers and unit commanders must be taught at Staff Colleges such as those at Camberley and Latimer. This fact is now fully accepted and the present day problem is one of ensuring that the subject is taught to good effect at the various establishments concerned. Peace-keeping should be regarded in exactly the same way and taught as a normal phase of military activity by those countries likely to become involved in it, which certainly includes the United Kingdom.

The purpose of mentioning the courses for company commanders run at the School of Military Intelligence in the early 1960's is to emphasize the fact that fighting subversion and insurgency was regarded as a specialist subject up until very recent times in this country, and the idea of treating it as a normal part of warfare is still foreign to some of the older officers in the army. Of course there are genuinely specialist aspects to this sort of war as there are to other sorts of war, and courses for specialists run at special establishments must take account of them. For example the syllabus of a Signals Officer's course must take account of the problems which may face an officer running a signals detachment in a counter-insurgency situation in the same way as it takes account of the problems which would face him if the Russians invaded Germany, and the same applies to Intelligence Officers for that matter. Furthermore there may be a requirement to run courses in some subjects which are only relevant to counter-insurgency operations such as tracking, in the same way as there is a requirement for running courses in subjects which are only relevant to conventional war such as the handling of nuclear artillery. The provision of an establishment to handle this requirement is considered in the next chapter but it no way mitigates against the requirement to bear in mind the needs of fighting insurgents and of carrying out peace-keeping operations when preparing the syllabus at all training establishments whether they cater for the needs of specialists or not.

At this point it might be helpful to look in rather more detail at the sort of education and training required at the various levels. If it is accepted that fighting insurgency and carrying out peace-keeping operations should be taught progressively as normal phases of military activity, it is necessary to ensure that officers get as firm a grounding in these subjects at the start of their careers as they do in other aspects of their profession. In terms of the British army this means that the Royal Military Academy at Sandhurst has the primary responsibility for teaching cadets the fundamental nature of insurgency and peace-keeping with particular reference to the political background. At the moment at Sandhurst, all cadets get some instruction in the nature of subversion and insurgency but the subject is only covered in detail by those who elect to study it as their special subject in preference to subjects such as 'A History of Strategic Thought' or 'Problems of Morale and Leadership'.[1]

Peace-keeping is not yet taught as a subject in its own right and although it is of lesser importance there is none the less a case for including the fundamentals of it in some part of the course. Those regular officers who are commissioned other than through Sandhurst can only study the fundamentals of these subjects, as they do other military subjects, by private study and the most that can be done is for their attention to be drawn to their responsibilities in this respect by their commanding officers in the early stages of their careers. In the British Army relatively few non-regular officers reach the position of commanding companies or squadrons, or of holding staff jobs concerned with operational policy, and therefore the theory of subversion, insurgency and peace-keeping is of less importance to them than the techniques which they should learn after joining their units. The problem of teaching the subject to these people academically can therefore be ignored, as indeed it must be because of the time available which is at any rate in line with practice relating the teaching of the theoretical aspects of other sorts of war.

[1] The Head of the War Studies Department at Sandhurst points out that over half of the cadets have chosen 'Guerilla and Revolutionary Warfare' as their elective subject since the option has been open to them. There are five subjects from which to choose.

The next phase of an officer's education is that which he gets within his unit. At this stage instruction is bound to depend largely on circumstances. For instance, if a battalion knows that it will shortly be moving to a country in order to take part in a specific counter-insurgency or peace-keeping operation, its training programme is sure to be geared to that particular campaign, and will almost certainly be heavily slanted towards teaching the actual techniques which have been found to be effective in the area concerned. On the other hand if the battalion is not so directly involved with a particular situation, training can and should be geared towards demonstrating and practising methods of approach which enable techniques to be fitted into the basic principles. For example training can be devised to teach the ideas contained in Chapter 6 and this is more likely to pay dividends in the long run than training centred round the techniques found useful in past campaigns. A few absolutely routine techniques will of course have to be taught to the soldiers in any case, such as the procedures used for handling riots or for cordoning streets or villages, but most of the officer training should be directed along the lines mentioned above. The correct method of approach can be taught either by cloth model studies or by full scale troop exercises.

If cadet training schools concern themselves largely with instruction about the fundamentals of the subject, and if units concentrate on teaching the right method of approach together with some of the more common techniques, it means that the rest of the educational and training requirement must be handled by arms schools and staff colleges. These establishments must also teach enough about the right method of approach and about techniques to enable students graduating from them to return to their units as instructors. Altogether this is a heavy commitment and unless there is effective co-ordination between the establishments concerned there will inevitably be gaps in an officer's knowledge.

In considering what should be taught in each school or college it is only necessary to bear in mind what the individual is being trained for on each course. For example the School of Infantry in teaching counter-insurgency to company commanders should concentrate on the correct handling of information because this is the essential concern of commanders at company and battalion level. At the same time some instruction should be given in the business of using non-military techniques to achieve operational aims because commanders at company and battalion level are likely to find themselves acting as military members of joint committees. On the other hand the Staff College is primarily concerned with training men to act as staff officers to senior commanders or to be senior commanders themselves in due course. The teaching should therefore concentrate on the problems of building up a campaign using civil and military methods. But senior commanders and their staffs must also understand what the commanders at lower levels should be trying to do so the teaching should include some reference to the tactical handling of information. The same considerations apply to peace-keeping. Arms schools should teach officers how to set about doing the task itself, while staff colleges should concentrate in explaining the background so that future staff officers in the Ministry of Defence, in overseas command or in a peace-keeping force can understand the problems facing their national contingent and provide the necessary support and direction for it.

The situation with regard to teaching the fundamentals to cadets at Sandhurst has already been mentioned and it is now necessary to say something about existing courses at British Arms Schools and Staff Colleges. In this connection it can be said at once that although much improvement to the present system still needs to be made if officers are to be adequately trained and educated, the situation is immeasurably better than it was ten years ago with one glaring exception. This concerns the instruction given at Arms Schools other than the School of Infantry with regard to preparing officers of the arms concerned to convert their units to infantry and use them in that role should it be necessary. It is true that units converted in an emergency often appear to put up a good performance after only a few weeks training in techniques appropriate to the particular situation, but although this reflects great credit on the intelligence and enthusiasm of the officers, and the versatility of the soldiers, it does not mean that the task is being done as well as it should be with particular reference to the long term implications. This fact is often overlooked because the unit concerned is unaware of what it should be achieving, and because senior officers themselves are sometimes unaware of it as well. Converted units are basically as untrained in this sort of warfare as infantry units were ten years ago, and a policy decision should be made either to maintain enough infantry battalions in the army to carry out all likely infantry tasks, or to give the officers of armoured and artillery units the same amount of training in the infantry aspects of counter-subversion and counter-insurgency operations as is given to the infantry officers themselves. Although the second course would be time consuming it would not be excessively expensive.

It will be convenient when considering the training given in countering subversion and insurgency at the main British Staff Colleges[1] and at the School of Infantry to look at the problem under three headings. First, is the content of the instruction given right and well balanced? Second, is there enough instructions at each level? Third, is it well co-ordinated so as to give a sound progressive education to officers throughout their careers? Assuming that the fundamentals are correctly taught to officers at the start of their career, the question of balance refers to the three broad areas of instruction previously mentioned, i.e. the co-ordinated build-up of a campaign using civil and military methods to achieve a given aim, the correct method of approach with regard to the tactical handling of information, and instruction in particular techniques. Looked at in this way it is immediately apparent that the weak link in the chain is the teaching of the correct tactical method of approach. Both wings of the Staff College and the School of Infantry at the Company Commander level go into considerable detail about the overall handling of campaigns, and all three establishments include instruction in the basic techniques either in isolation as in the Platoon Commanders' course at the School of Infantry, or as part of the other instruction at higher levels. The correct method of approach is not taught because it is not fully understood, although at the Staff College at least there is a realization that a gap exists, and that it is concerned in some way with the function of intelligence or the handling of information.[2] A much less serious weakness with regard to content and balance concerns the emphasis placed on one part of the operational spectrum as opposed to another. Thus, thanks largely to the current influence of Viet Nam, Hong Kong and Northern Ireland,

[1] The Joint Services Staff College, The Staff College at Camberley, and The Junior Wing of the Staff College at Warminster.
[2] Interview given to author by members of the Directing Staff, 3 June, 1970.

teaching tends to be polarized towards the extremes, that is to say the handling of riots on the one hand and the countering of major guerilla forces on the other. As a result the countering of small terrorist groups especially in urban areas gets less than its fair share of attention.

It is more difficult to express an opinion on the question of whether enough training time is devoted to the fighting of insurgency and subversion as opposed to other forms of warfare, because so many factors beyond the immediate scope of this study are involved such as the relative complexity of teaching people to fight one sort of war as opposed to another. For example, even if it was agreed that preparing to counter insurgency should take priority over all other military activities, it might still be unnecessary to allocate more training time to it than to preparing for some other aspect of warfare which has a lower priority, but which was much more difficult to teach. Little value can therefore be gained from working out the proportion of a course devoted to the study of this sort of war as opposed to other sorts of war, and the only useful assessment is whether the total time allotted is sufficient for achieving the aims of producing officers who are well enough trained to carry out the tasks which they must be prepared to undertake. This in turn is impossible to assess other than in terms of the opinion of individuals who have knowledge of the results achieved in the various operational theatres. One fact is, however, apparent from an examination of the training carried out in the schools and colleges under discussion and that is that the proportion of the total time allocated to the study of this sort of war goes down as the seniority of the student goes up. If it could be shown that over a period the performance of junior officers was noticeably more impressive than that of senior commanders and their staffs, it might indicate that more attention should be paid to instruction at the higher levels. Although proof as such is not of course available, it could be argued that the record of past campaigns indicates this, and it is probably fair to conclude that more training is required for those who may be employed in counter-insurgency operations above the level of a company commander.

In this connection the point is sometimes made that the Directing Staff of the Colleges concerned have little control in the matter of how much time should be devoted to studying the countering of subversion and insurgency as opposed to other forms of war, because their teaching priorities have to be geared closely to National Defence Policy as laid down by the Government. If this is the case there is a strong case for altering the system for two reasons. First, although the provision of manpower, units and equipment has to be geared very closely to National Defence Policy because of the expense involved, it has to be admitted that those who lay down the policy may make mistakes and the small amount of extra money needed to ensure that officers study what would happen if the government's predictions turn out to be wrong should be regarded as money well spent. Indeed it can be said that such study, which is a genuine form of preparation, has its part to play in deterring potential enemies and in encouraging potential allies even if it carries less weight than more material forms of preparation. The second reason for altering the system is that even if the likelihood of having to fight insurgency is considered to be less than that of having to fight other sorts of war, the need to prepare for it is indisputable. But preparation involves far more than training people who may have to go to the scene and deal with trouble when it arises. It is just as important that all those concerned with organizing and administering the army should have a thorough understanding of what is involved so that they can play their part to good effect. The significance of continuity in fighting insurgency as described in Chapter 6 affords a very good example of this. To provide continuity in a theatre of operations involves taking all sorts of measures with regard to the deploying of units, the posting of men and even of the composition of the overall order of battle. It has ramifications in the administrative field with regard to the amount of unaccompaniment which soldiers may have to undergo which affects such matters as recruitment, pay and accommodation. In practice much has to be sacrificed in order to provide continuity and the right decisions are only likely to be taken throughout the various headquarters and in the Ministry of Defence, when all officers in them understand what is involved. For both of these reasons it is important that the subject should receive adequate coverage in schools and colleges and that decisions regarding defence priorities should not be translated too literally into hours devoted to one subject rather than another.

The next point to consider concerns the co-ordination between the various schools and colleges which is required for ensuring that the training and education of an officer progresses satisfactorily throughout his career. Whether by accident or design a measure of co-ordination appears to have been achieved, and although there is some duplication at each level this is probably no more than is necessary for the essential purposes of revision and for ensuring that officers who have missed instruction at a lower level get enough of a grounding at the higher level to enable them to absorb the main instruction given there. The only exception to this situation concerns the training given at the Joint Services Staff College where officers of the Royal Navy and the Royal Air Force are brought into the picture for the first time. It may be that with the disappearance of unified overseas commands there is no need to teach even the rudiments of subversion and insurgency to officers of these Services but such people may be appointed to positions of influence, if only as staff officers in the Ministry of Defence, and it would therefore seem that a case exists for recasting the instruction at the Joint Services Staff College to take account of this requirement. This is not to suggest that no instruction on the subject is included in the present course at the Joint Services Staff College but it is certainly insufficient for giving a reasonable grounding in the subject to officers who may not have studied it at all in the past. These people do not need to be taught any of the operational techniques involved, nor do they need to know much about the method of handling information in order to bring troops into contact with enemy groups. What they do need to understand is the essential nature of insurgency with particular reference to political considerations, the relationship which should exist between the military and civil authorities, and enough about tactical procedures to enable them to appreciate the advantage and disadvantages of using particular methods and weapon systems.

But although training in counter-subversion and counter-insurgency is at least attempted at every level, the same cannot be said for peace-keeping. Only at the Staff College in Camberley is reasonable provision made for covering the subject. At other schools and colleges it is hardly covered at all, instruction being limited to one or two lectures by guest speakers. In this respect the British Army is lagging well behind the armies of a number of other countries which include peace-keeping in the syllabus of their National Staff Colleges,[1] Canada, for example, runs three two-day exercises at their equivalent of the British Joint Services Staff College, and similar emphasis is placed on the subject in the training of more junior staff officers.[2] Against this it must be admitted that the United States Army scarcely touches on the subject in its teaching at any level, which is unfortunate because the time may come when it will be obliged to employ soldiers in this role. But irrespective of the situation prevailing in other countries there is no doubt that the position with regard to the British Army is unsatisfactory. Instruction in the fundamentals of peace-keeping and in the techniques which have been found useful in the past should be given at every level.

A further point which is closely connected with education and training is worth mentioning in this chapter. This concerns consultation with allies over methods of fighting subversion and insurgency in areas where British troops might become involved in helping those of another country, or where British and allied troops together might go to the assistance of a third country. Unless some consultation had taken place in advance it is almost inevitable that misunderstandings would occur especially as some of the contingents in a joint force would have no practical experience of any sort. Bi-lateral consultations and possibly some joint training would cost very little, but might make a significant contribution towards increasing the effectiveness of subsequent intervention.

One of the most important by-products of proper training concerns the selection of officers to fill the more important posts in counter-insurgency and peace-keeping operations. In the days when there was virtually no training in these subjects, officers had to be selected either because they had gained previous operational experience or because they were thought to be men of such high calibre that they would instinctively know what to do when confronted by the problems on the ground. This put an unwarranted responsibility on to Selection Boards and the Military Secretary's department, and in many cases the selectors themselves had neither the training nor the experience to make good

[1] A. J. WILSON, *Some Principles for Peace-Keeping Operations, A Guide for Senior Officers*, IPKO Publications, 1967, p.10.
[2] Information supplied by the Commandant of the Canadian Armed Forces College.

choices. It is hardly surprising that unfortunate selections were made from time to time, but it is none the less true that some of the failures, or delays in achieving success, suffered by the British Army in the twenty years following the Second World War were due to this cause. By ensuring that all officers are properly trained at each level and that qualification for promotion includes qualification in this field as in other ones, selection boards will be able to revert to their proper function of choosing the best person for a particular task from out of a number of qualified men.

It may be felt that altogether too much has been made in this chapter of the need to educate officers to take part in counter-subversion and counter-insurgency operations. Those who think along these lines may be interested to compare the training given to British officers with that undergone by officers of the United States Army, bearing in mind that both armies are obliged to maintain a conventional and a counter-insurgency potential. In conducting this comparison two lines of approach suggest themselves. The first of these is to look at the amount and type of training carried out by the ordinary officer throughout his career and the second is to consider the extent to which specialist training is given in subjects closely connected with counter-insurgency operations. But before looking at these two aspects it is necessary to say a word about the influence which the war in Viet Nam has had on training in the United States Army. Naturally the requirement to provide a steady flow of men to fill appointments in Viet Nam has greatly influenced the education now being given to officers at every level, but it is none the less true to say that this is not the whole story. More interesting are the steps which are being taken to ensure that any future campaign is handled efficiently from the start, with particular reference to the measures necessary for helping an ally. Training of individuals to fight in Viet Nam has to go on, but an encouraging amount of thought, especially at the higher levels, is directed towards evolving procedures, for use in other parts of the world in the future.

The United States Army, like the British Army, bases its policy for educating officers in counter-insurgency on the need to include the subject in the syllabus of arms schools and staff colleges at every level. Where it differs from the British Army is in the amount of time which it devotes to the purpose and in the detailed content of its instruction. It is not easy to make an accurate comparison between the two armies in terms of time allocated, because it is difficult to interpret entries in a syllabus properly. For example, a subject under the general heading of counter-insurgency may be shown as taking up so many hours of training time, but when examined closely it turns out that the matter being taught is a straightforward administrative or movement problem which has been included under the heading of counter-insurgency merely because the exercise in which it appears is one with a counter-insurgency setting. Conversely, instruction on 'The Battalion in the Air Mobile Role' for example might appear in the syllabus under 'Aerial Employment' although the subject matter might be exclusively concerned with counter-insurgency operations. But despite these difficulties there is no doubt whatsoever that the United States Army devotes more time to the subject than does the British Army. It would probably be fair to say that American officers below the rank of Lieutenant-Colonel spend at least twice as long studying the subject as their British counterparts; above that rank the ratio goes up even further. So far as the detailed content of the instruction is concerned there appears to be less difference between the position of the two armies, the weak point in both cases being in the area of the tactical direction of the troops. Both armies tend to concentrate on the civil military background of operations, although the Americans show more interest in the fundamentals of the subject and the study of past campaigns. Both armies also understand field techniques well enough, although the British are stronger in the area of low level counter-insurgency operations at the moment. Certainly the American Army is far more closely geared to operating in an advisory capacity but this is a subject which is more directly related to the training of specialists.

Earlier in this chapter[1] mention was made of the problems involved in training specialists such as signals officers or intelligence officers, and the point was then made that the particular needs of fighting insurgents should be borne in mind when courses were being prepared at whatever school was giving the instruction. It was also said that there might be a requirement to run courses in some subjects which were only relevant to counter-insurgency operations such as tracking. The United States Army

[1] See p.169 above.

approaches the problem in a radically different way to the British by concentrating in one large military post all specialist courses relevant to counter-insurgency operations. In other words Military Advisers, Psychological Operations Officers, Civil Affairs Officers, Special Forces Teams, Intelligence Officers destined for employment in counter-insurgency theatres, and various other specialists are all trained in one institution by instructors who owe allegiance to one commander. Furthermore this officer also commands all operational Special Force units in the Continental United States. By this arrangement a high degree of cross-fertilization is achieved amongst a large number of high class people who specialize in fighting insurgents. Not only does this mean that the actual courses are more efficient than they would be if they were dotted around schools and establishments throughout the country, but it also means that there is a considerable 'spin off' of expertise which is carefully collected and analysed by a research group of Combat Development Command consisting of over fifty officers and civilians using modern methods of data storage and sited on the same post. This group not only feeds the result of its work back to the instructors on the courses, but it also passes it to arms schools and staff colleges throughout the army so that the latest ideas are available wherever they are wanted. The cumulative influence on the development of ideas exercised by this complex of highly organized specialists is far more powerful than anything which the British Army can produce.

In concluding this chapter it may be said that although the training and education given to officers of the British Army with regard to peace-keeping and the countering of subversion and insurgency is much better than it was ten years ago, it is still far from adequate. The most urgent requirement is to arrange for officers to be taught the correct method of approach with regard to the tactical handling of information in counter-insurgency operations, and the training carried out in units, schools and colleges needs to be re-cast around this requirement. In passing it might not be out of place to suggest that an understanding of this idea would be of help to the United States Army as well. Other necessary alterations to the present system include generally increasing the time allocated to teaching counter-subversion and counter-insurgency in schools and colleges throughout the army, particularly in those which cater for more senior officers; ensuring that officers of arms other than infantry get proper training if it is intended to continue converting such units to the infantry role; and finally, greatly increasing the training given in peace-keeping operations so that all officers have a sound knowledge of this activity.

Chapter 10

Provision of Units, Specialists and Equipment

Experience over the past twenty-five years regarding the provision of units for counter-insurgency or peace-keeping operations shows that trouble almost always breaks out in areas where it is not expected, that the initial requirement for units is often greater than can be provided from whatever strategic reserve is held for dealing with the unexpected, and that the requirement lasts for longer than most people consider possible in the first instance, although the intensity of operations may fluctuate and the requirement for units decline as time goes on. The disturbances which broke out in Cyprus at the end of 1963 afford a perfect example of these three factors working together. The trouble had not been foreseen in the sense that no specific provision had been made in the Order of Battle for providing units to deal with it; the initial requirement for troops was greater than could be found from the strategic reserve so that units had to be diverted from other tasks, e.g. initially the battalion assigned to the Allied Mobile Force and subsequently battalions serving in the British Army of the Rhine; and the commitment has continued, albeit in a reduced form, for far longer than was originally foreseen. One or more of the three factors has been relevant to nearly every emergency which has taken place since the end of the Second World War and it looks as if Northern Ireland may afford another example of all three of them operating together. The most important difference between the situation which exists now and that which existed twenty years ago, is that the overall size of the army is much smaller so that although some provision is still made in the Order of Battle for maintaining units to deal with the unexpected, the scope for borrowing those committed elsewhere is greatly reduced.

It is against this background that the question of providing units to take part in counter-insurgency and peace-keeping operations has to be considered. Broadly speaking there are four ways in which units can be made available. First the existing system can be continued in which case each new situation will have to be handled by calling on troops committed to other tasks as soon as the small reserve which is kept specifically for the unexpected is used up. The second way is to convert strategic reserve units of a type not required by the situation into a type that is required. In practical terms this has meant using artillery or armoured corps units as infantry, but this might not always be the case in the future. The third way of dealing with the situation is to raise new units when a particular demand arises, and the fourth is to maintain a higher reserve for dealing with the unexpected in the first place, which in effect means keeping a larger army. The advantages and disadvantages of these four courses are discussed below.

The advantage of relying on the system whereby units which are carrying out a role in one place, are switched to dealing with an emergency elsewhere should one arise, is that it is on the surface the most economic method of handling the situation. By using this method the army can be kept as small as possible, and when nothing is going on, no resources are being wasted. There are however serious disadvantages. If for example a battalion from the British Army of the Rhine is sent to Northern Ireland it not only leaves a gap in Germany but it quickly loses its mechanized skills, and its absence will have an adverse effect on the training of the formation from which it is taken. Furthermore the married men are obliged to leave their families in some German town which is considerably less satisfactory than leaving them in England where they can at least pass the time with their relatives and friends. For all these reasons the unit's tours in the emergency area will have to be restricted to a few months which is thoroughly unsatisfactory from the point of view of continuity, in addition to which it means that another unit has to be uprooted to relieve it unless the commitment has come to an end in the meanwhile. This in turn means further dislocation and disturbance, a drop in the efficiency of the formation from which the new unit is drawn and a lowering of the new unit's own standards in its normal role. In the end the emergency is conducted badly, unaccompaniment increases, units become less efficient and men become discontented, with the result that they leave the service or at any rate fail to re-engage when their current engagement expires. The unit is then undermanned which puts an increased load on those remaining with further adverse consequences.

The next alternative to consider is the converting of units from one role to another. Once again the advantage is that on the face of it at any rate, the system is more economic than keeping a larger army or raising extra units. But there are a number of disadvantages involved. In the first place neither the officers nor the soldiers of arms other than infantry are properly trained to act as infantry, a matter already mentioned in the last chapter. Another important disadvantage is that armoured and artillery units contain a high proportion of specialists such as mechanics, electricians, drivers, wireless operators and tank gunners whose expensive skills are largely wasted if they are used for patrolling through the jungle or for controlling crowds in the streets of some large city. Not only are these skills wasted under such circumstances but they also rapidly deteriorate so that tours in the converted role have to be kept short. Finally armoured and artillery units are weaker in manpower than infantry battalions which means that a converted unit cannot take a commitment from an infantry battalion on a one for one basis. In fact it is necessary to put two artillery units out of action in order to make one infantry unit of equivalent strength.

By comparison with either of the two alternatives so far considered, the advantages of raising extra units to deal with an emergency are very great. From an operational point of view a battalion can be raised and trained to the extent necessary for it to carry out routine duties in the United Kingdom or in Germany within about nine months thereby freeing another battalion to take part in the emergency. Within a further nine to twelve months it could be ready to take part itself. But although there are less operational problems involved in raising extra units, there are considerable financial and administrative difficulties. There might even be a major political drawback to this course of action, because it involves admitting that the army which is supposedly capable of defending the country from a major assault, is not able without being enlarged to handle what by comparison must appear to be a minor affair. How great this political difficulty would be, would depend largely on the circumstances. If the emergency arose at a time when the army was being reduced, it would only be necessary to delay the run down which might pass virtually unnoticed. If the emergency arose following a change of government, the expansion could be handled by a reference to the previous government's miscalculation in maintaining too small an army. If on the other hand the government had no one to thank but

184

itself for the number of units in existence, the problem of raising more might be a grave one from a political point of view. In any case experience gained during the whole of the period since the end of conscription indicates that there would be difficulty in finding the necessary manpower to raise additional units. There would also be the problem of disposing of the units after a year or two once the emergency resolved itself.

The fourth alternative, i.e. that of maintaining a sufficiently large strategic reserve, is obviously the most satisfactory from an operational and administrative point of view. There are however three disadvantages which merit consideration. The first and least important is that if unexpected commitments fail to materialize those units serving in the strategic reserve might not have enough to do. Ever since the end of the confrontation with Indonesia in Borneo and the withdrawal from Aden, there have been those who have felt that in the long term the army was likely to suffer more from having too little to do than too much, and although the problem has not yet arisen it could do so by the middle of the decade. The second disadvantage is that a larger strategic reserve can only be maintained if more money is available for the purpose, and this might have to be found to the detriment of some other military capability or at the expense of one of the other services. The third disadvantage is that the army is already under-recruited and it might not be possible to find the manpower necessary for maintaining more units without reintroducing some form of compulsory service. Against this it can be said that ever since the end of National Service the army has always been a few thousand under strength, and the percentage of undermanning has been much the same whether the target was 190,000, 180,000 or 165,000. It is perfectly possible that the percentage would remain roughly the same if the target was raised to 220,000 or reduced to 150,000 the reason being that the Government works out as accurately as possible the pay and amenities needed for recruiting the numbers required, and unless there is a small shortfall it can never be certain that it is not offering more than is necessary. Indeed it can be argued that it would be easier to maintain a slightly larger army than that catered for under present plans because its members would not be so heavily overemployed and might therefore be more contented.

Although the most pressing problem during the past ten

years with regard to the provision of units for peace-keeping and counter-insurgency operations has been to find infantry battalions, a similar one exists with regard to logistic units. There are even fewer of these available than there are infantry battalions, and proposals to increase the numbers are even less popular because of the reluctance of soldiers and politicians alike to enlarge the so-called tail at the expense of the teeth. But in fact the analogy is a bad one. The logistic units represent gums rather than tail, and without gums the teeth won't bite. This aspect of the problem is accentuated by virtue of the fact that for certain United Nations Operations Britain might be asked to provide logistic units alone. Some provision has been made for this in the form of logistic units retained in the Reserve Army specifically designed to take part in peace-keeping operations or to relieve other units who could carry out the task. In addition there is some scope for converting artillery and armoured corps units into certain sorts of logistic units such as transport squadrons should the need arise, and a conversion of this nature would have less drawbacks than converting them to infantry. But taken overall there is likely to be as serious a shortage of logistic units by the middle of the 1970's as there will be of infantry battalions.

It is not possible to make a firm recommendation as to which of the four courses should be adopted for providing the required number of units to take part in counter-insurgency and peace-keeping operations: in any case the actual plan at any given moment is likely to contain elements of two or more of these courses. Whereas it is unquestionably better to maintain a sufficient number of properly trained infantry battalions than it is to adopt *ad hoc* methods of providing them at the eleventh hour, it is equally certain that no good purpose would be served in the long term by having them if it meant sacrificing the ability to take part in more sophisticated forms of warfare unless it becomes apparent that such wars are no longer likely to occur. The balance between one operational capability and another can only be decided at the highest level in the Ministry of Defence after consideration of all the known factors, in the same way as the balance between military and civil expenditure can only be taken by the Government. A decision to maintain even six or eight more battalions than are currently considered necessary would require an appreciable amount of money in relation to the total Defence Budget at a time when the Government is trying to cut its overall expenditure, and it may well be that it would insist on the money being found at the expense of some other Defence capability. It is none the less right and necessary in a book on this subject to emphasize the factors relevant to peace-keeping and counter-insurgency which those responsible for taking the decisions mentioned above should bear in mind. In particular the great importance of continuity and of deploying properly trained units must be stressed, not least because operations carried out on any other basis produce results which are very expensive and politically dangerous.

By comparison with holding extra major units in the army's Order of Battle the rest of the subjects discussed in this chapter are relatively cheap, one of the least expensive being the need to keep enough specialist individuals and units within the army to enable essential civil services to be maintained in the event of civilians being unable or unwilling to maintain them. At the end of the Second World War the army contained individuals or specialist units capable of carrying out all sorts of civilian functions.For example, it could run ports, railways, power stations and sewage works as well as supervise the operation of mines and many types of industrial plant. As the army has contracted, many of these capabilities have been sacrificed in order that remaining resources can be concentrated in the most important areas, and the process has been accelerated in the past five years as a result of the recent defence cuts. Once again it is not possible to make specific recommendations as to the exact specialist posts and units which should remain within the army, but it is necessary to suggest that the greatest care should be taken before any more cuts are made in this direction. Should the army be required to take part in peace-keeping or counter-insurgency operations outside the United Kingdom it is possible that some specialists of the sort mentioned would be required. Even within the United Kingdom a situation might arise in which the army was required to provide men for this purpose.

One form of specialist activity provision for which requires discussion in more detail, concerns the ability of a government to disseminate its views and policies in an advantageous way in a situation of subversion and insurgency. This matter was considered in some detail in Chapter 4 where it was stated that the army should be prepared to provide staff officers trained in the techniques of psychological operations to act as advisers to security committees or commanders, and also teams capable of carrying out such measures in this field as may be considered necessary. In Chapter 4 the point was made that although there is no particular reason why the army should be responsible in the sense that properly trained civilians would be just as good at doing the job itself, the army is the only organization which can train and hold the right sort of people equipped with the right sort of equipment in advance of a situation arising. In practice the Psychological Operations resources controlled by the Services at the time of writing are very limited. There are 'PSYOPS' staff officers at three headquarters overseas, and there is also one in the Ministry of Defence. There are also two officers on the strength of the Joint Warfare Establishment who run short courses on the subject and those who have attended are earmarked as being suitable for employment in this capacity should the need arise. So far as service manned teams are concerned there is only one in existence which consists of an officer and eleven men. This is operating at the moment and is subsidized by the government of the territory in which it is deployed. A small number of civilian teams are also being raised for use by the same government in its own territory. The only reserve team is one which is being formed as part of an infantry battalion. If required for an operation it would have to be divorced from its parent unit which therefore has all the inconvenience and work of raising and training it without any prospect of benefiting from the arrangement. No unit relishes the prospect of losing an officer and eleven good men at short notice.

In terms of the situation envisaged in Chapters 4–7 it can be seen that the total Psychological Operations resources available would hardly be sufficient to provide an adequate staff in the headquarters of a Director of Operations let alone in provinces, counties or districts. Furthermore it would be quite impossible to provide any reasonable coverage of teams at short notice. There is certainly a case for holding a few staff officers and one or two teams at short notice as part of the strategic reserve so that this essential aspect of counter-subversion and counter-insurgency can be covered as soon as troops are despatched. Ideally staff advisers and teams should be sent well in advance of any troop deployment at the same time that reinforcements are sent to the intelligence organization, but this is often impossible because the authorities at the receiving end are unlikely to understand what is required in this direction until the arrival of the army. But to maintain a capability of this sort would involve diverting resources from elsewhere and although the financial and manpower requirement is small in relation to maintaining extra infantry battalions it is none the less significant. A further point is that few of those responsible for running campaigns of counter-insurgency understand psychological operations so that until senior officers are fully educated in this respect the resources would probably be left unused or, at best, misused. But it must be admitted that the present situation is not satisfactory and a move should be made to improve the situation. Although the British seem to persist in thinking of psychological operations as being something from the realms of science fiction, it has for many years been regarded as a necessary and respectable form of war by most of our allies as well as virtually all of our potential enemies. Some evidence of this fact can be adduced by comparing the two instructors, four staff officers and twelve team members employed by the British, a total of eighteen men, with the numbers maintained by some of our allies. For example the West Germans maintain psychological operations units totalling about 3,000 men in their regular army to say nothing of reserve army units. The Greeks and the Turks each employ around 300 men and the Italians include one full strength Psychological Warfare company in their Order of Battle. The Americans maintain a psychological operations capability in all their Joint Commands overseas as well as Psychological Operations Battalions in the Continental United States, grouped with Special Force Units and held ready for immediate deployment as required. Undoubtedly the British are 'bringing up the rear' in this important aspect of contemporary war.

An even more important subject than providing the capability to carry out psychological operations concerns the steps which should be taken for ensuring that the army is capable of reinforcing local intelligence organizations or setting one up altogether where none exists. Closely allied to this are the preparations which are needed for ensuring that men are held trained and ready to go to a theatre of operations and develop information by special methods as described in Chapter 6, either as part of the intelligence organizations or as a Special Force closely allied to it.

So far as the British Army is concerned the present situation with regard to providing men to reinforce or build up an intelligence organization is as follows. If no local intelligence organization exists the army can do no more than provide an improvised party after a long delay. Although some officers have the necessary experience they are not held ready to move at short notice, and even if they can be spared from their current jobs time is needed for them to be extricated and prepared. Naturally under such conditions there can be no question of their being trained together nor will they be accustomed to working with each other when they arrive at their destination. If a local intelligence organization does exist there are people trained and ready to reinforce it in the form of Military Intelligence Officers but there are very few of them and only two, together with a few sergeant interrogators, are held at short notice to move. There are in addition a number of Military Intelligence Liaison Officers held on the staffs of certain headquarters but they are intended merely to act as liaison officers between the military headquarters concerned and the local intelligence organization. All the same they have some knowledge of the intelligence world and could be used as reinforcement Military Intelligence Officers if necessary. No provision is made for setting up a force trained to carry out the sort of special operations required for developing information in close connection with the intelligence organization.

In theory at any rate this situation is extremely unsatisfactory because speed is of the utmost importance in getting the intelligence organization going properly and in preparing a force to carry out special operations. Insurgents are particularly vulnerable in the early stages of a campaign because at that time they have not perfected their security measures and may not have cajoled or terrorized more than a relatively small proportion of the population into supporting them. If the government is able to develop its full potential quickly and mount effective operations in the early stages of the uprising it has a chance of cutting years off the time for concluding the business successfully or of avoiding defeat. In practice the situation is not quite as bad as it looks because in some troubled areas the authorities are reluctant to ask the army for reinforcements for their local intelligence organization, and while they are being persuaded of the necessity for doing so, Military Intelligence Officers can be assembled from a reserve of those already trained. But this takes no account of the sort of situation in which troops are deployed rapidly and unexpectedly into an area where no intelligence organization exists nor does it reduce the need to hold men trained and ready to set up a force to carry out Special Operations.

An effective way of dealing with this problem would be to establish a unit which could carry out the two separate functions of setting up or reinforcing the intelligence organization and of providing men trained in operations designed to develop information by special means. If a unit of this kind were formed the element designed to set up or reinforce the intelligence organization would consist of a number of officers available to move at short notice when needed. These men would be majors or captains and they would be backed by a number of other ranks to act as drivers and clerks. The unit could be a relatively large one in which case there might be three of four groups each consisting of a major and several captains, the major being intended for deployment to a provincial or county intelligence headquarters, and the captains to districts: a unit of this size would be commanded by a lieutenant colonel or senior major who could deploy to the intelligence headquarters of the country concerned. Alternatively the unit could be much smaller and consist of only one group in which case the major would be intended for deployment to the intelligence headquarters of the country concerned and the captains to provinces or counties. In the second case the intention would be for the group to make a start on tackling the problem and then fit subsequent reinforcements into the lower levels as they could be found. Prior to being committed the officers would be engaged in training which should ensure that they had reached a high standard before starting to operate. Not only could they study the business of collecting background information but in addition each officer or group of officers could specialize in a particular area of the world. Thus one group or individual might specialize in Scandinavia, another in the Mediterranean countries, a third in Africa and a fourth in Asia. This specialization would not be designed to preclude deployment in other parts of the world but it would enable the unit as a whole to build up an overall knowledge of likely areas. Specialization should involve visits and some elementary examination of the languages of the area as well as a thorough study of the area's problems.

The provision of men trained in operations designed to develop information by special means produces a different problem because the teams which actually operate on the ground are bound to consist mainly of local people such as surrendered or captured insurgents. What is required before the start of a campaign is a cadre of men who have studied the various methods used in the past, and who can go at short notice to build up teams in the country concerned. On the whole they should be younger men than the officers held for reinforcing or setting up an intelligence organization and such parts of their training as is not concerned with the study of past methods should be slanted towards living in 'the bush' and the use of various weapons, as opposed to means of collecting, processing and distributing information. At the same time a knowledge of intelligence techniques would be useful to them, and they could also specialize in particular geographical areas. In fact the closer the link between the two parts of the unit the better because it may well turn out that when deployed, the special operations cadre will be working in close support, or even under command of the people collecting background information. The actual organization of this cadre must be geared to the fact that once deployed the men in it will be used to direct indigenous teams rather than to operate themselves. On this assumption it should be in a position to provide a number of cells each consisting of an officer and one or two training sergeants. Ideally there should be enough cells available so that one can go to each district but it may be that the trouble has broken out in a country which is too big for this to be possible. For the purpose of this study it is suggested that the cadre should consist of six or eight cells.

If a unit were to be raised on the lines suggested, it is worth considering whether it might not also fulfil certain other functions which are relevant to counter-insurgency operations, and which are not at this moment covered by any military organization. One obvious gap which the unit could fill concerns the research and development of equipment which might be of value to the army in fighting insurgents, such as special weapons, communications systems, data storage devices for use in the collection and development of information, and other technological advances of a similar nature. Another gaping void which the unit might fill, lies in the development of ideas for use by specialist or orthodox units in counter-subversion or counter-insurgency situations. If a carefully selected tactical development section were to be included in the unit's establishment, it should be able to come up with some useful ideas using a two pronged approach involving the examination of records on the one hand combined with discussions with the units officers on the other, all of whom would be specialists in this field. The output of those working on equipment research and on the development of tactical ideas would be passed into the bloodstream of the army as a whole via Arms Directors or the Ministry of Defence as appropriate.

Of all the specialist activities relevant to the prosecution of a counter-insurgency campaign none is more important than the provision of trackers. There is no truth in the theory that a man can only track in the country in which he has been brought up, but obviously he will only be able to do so effectively outside his own area if the terrain is similar to that in which he has been raised. Unfortunately it is practically impossible to train a man to track in an acceptable length of time if he has no experience of it, and it is equally unfortunate that no indigenous trackers exist in many of the countries in which counter-insurgency operations may be expected to take place. The problem therefore consists of identifying the countries which do have a sizeable population of effective trackers and equating them with possible areas of subversion. For example some North American Indians still track for a living, and they may be the only people who can track effectively in the cold, wet forests of the Northern hemisphere. On the other hand a number of African tribes still live by tracking in tropical forests and the same can be said of tribesmen in certain parts of South East Asia. Men who can track well in very arid conditions are far less easy to find but they do exist in certain parts of Africa. The task of the unit with regard to the provision of trackers might therefore fall into three parts. First it could carry out a study in order to decide where to get trackers from to meet the requirement of every likely campaign. Next it could make contacts in the countries from where the trackers would come in order to ensure that they can be collected quickly. This would involve selecting the right men and ensuring that the government of the country concerned was prepared to let them go. Finally it could maintain a small pool of trackers from different parts of the world who could be deployed with the first troops into a theatre and bridge the gap while others were being obtained from their native lands. Before

o

being committed these people could be used to train the men of the Special Operations teams in some elementary techniques and even more elementary training might be given to individuals from units so that they would at least know what to expect from good trackers when such people eventually arrive.

Finally there would have to be a small training section designed to instruct new members of the unit how to operate. In practice this might not consist of anything more than an officer in the unit headquarters with some clerical assistance, because most of the actual instruction would be given by other members of the unit, e.g. the group commanders.

At this point it is worth seeing what a unit such as that described would look like, and an organizational chart made on the assumption that the unit would consist of three groups designed to reinforce an intelligence organization and eight cells for activating special teams is given at Figure 6. A unit constructed in this way would certainly find itself concerned with a wide range of activities, but this itself is not uncommon in the army. Where it would be unusual to a greater extent would be in the way in which it would deploy. In effect it would not deploy as a unit but bits and pieces of it would go off to carry out different tasks in different places, while the part concerned with developing ideas and equipment would stay where it was. For this reason it might be felt that a unit so constituted was not a unit at all, and that its functions should be tacked on to one of the Arms Schools or possibly to the Defence Intelligence Centre. The objection to attaching such a growth to one of the Arms Schools is, first of all, that no arm has a monopoly of this form of war. The next objection is that the Arms School concerned is bound to be at least as much concerned with other forms of warfare, and the influential appointments in it might be held by people who had little experience of counter-insurgency operations. Furthermore when a conflict of interest arose for the allocation of resources as between this unit and the rest of the School, the unit might not get its fair share. A similar series of objections arises in connection with the possibility of attaching the unit to the Defence Intelligence Centre in that the Centre would be more interested in the wider aspects of intelligence than it would in a unit which is partly concerned with offensive operations. Here the problem of personalities would be even more acute because the sort of person who has grown senior

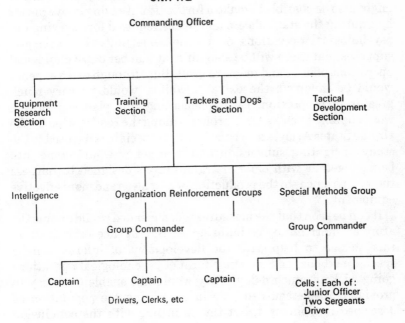

FIGURE 6

SPECIAL UNIT

UNIT HQ
Commanding Officer

Equipment Research Section · Training Section · Trackers and Dogs Section · Tactical Development Section

Intelligence · Organization Reinforcement Groups · Special Methods Group

Group Commander

Captain · Captain · Captain

Drivers, Clerks, etc

Group Commander

Cells: Each of:
Junior Officer
Two Sergeants
Driver

in the field of pure intelligence is likely to be totally unsuited to the world of special operations. At the same time there is a lot to be said for taking the unit on to some existing organization and it is suggested that an examination might be made into the merits of the various available alternatives, if only to save overheads.

It can not be denied that a unit of the sort described would be expensive even if overheads were shared. Some of the expense would automatically be offset by the disbandment of the Military Intelligence Officers because the responsibility for re-inforcing local intelligence organizations would fall on the new unit. It might also be possible to make a further small saving in overheads by grafting the staff officers and teams required for carrying out psychological operations on to the same units, but this pre-supposes that there will be a significant number of psychological operations officers and teams to graft. But although extra expense would be involved, the extra capability would be very much greater in proportion. By implementing this plan the British would be able to make some progress along the road plotted by the United States Army, and a community of specialists devoted to the study of fighting subversion and insurgency would come into being together with embryo organizations designed to harness their experiences to the production of new ideas and more effective equipment.

It can be seen that the measures recommended would enable the army at short notice to build up an intelligence organization, raise teams to help with the development of information by special methods, and get some form of psychological operations going without much delay. They would not enable the army to provide trained teams to help in organizing the population as described in Chapter 4, but this is in line with the conclusion reached in that chapter, to the effect that in the light of existing financial and manpower pressures, the situation does not warrant the holding of men ready in advance to do this job, because there would be time to select and train people after the subversion was discovered and before it was politically possible to deploy them.[1] But there is no doubt that it would be desirable to have such teams ready, together with military advisers and civil affairs experts. The training and holding of such people constitutes a second phase of preparation and one which the British Army may be

[1] See p.80 above.

obliged to implement if the threat of subversion increases. The United States Army already trains and holds a large number of people of this sort, having largely switched their special forces into this field and it would be possible for the British Army to do the same thing if the need arose.

Chapter 11

Conclusions and Afterthoughts

The stated purpose of this book was to draw attention to the steps which should be taken now in order to make the army ready to counter subversion and insurgency, and to take part in peace-keeping operations in the second half of the 1970's, and it can be seen that the recommendations fall under two main headings. The first of these, i.e. training and education, covers the measures which should be taken to ensure that officers take advantage of what has been discovered with particular reference to the tactical handling of information. The second covers the organizational measures required for ensuring that the right sort of units are available and that they are armed and equipped to the best advantage. It can also be seen that measures under these two headings interact on each other to some extent, in that the organizational suggestion for setting up a unit designed to concentrate such specialists as the British Army holds, would if implemented, lead to a better understanding of the problems of fighting insurgency and subversion, which would be important in terms of education and training.

It is difficult to know whether any particular recommendation deserves to be stressed in this concluding chapter, but there is certainly a case for underlining the proposal to set up the special unit mentioned, with all that it would imply in terms of being ready to intervene effectively, in the early stages of a campaign, and also in terms of the increased awareness of the problems of subversion and insurgency which it would generate. On the other hand it might be considered that disseminating the idea of developing background information into contact information was more important because it fills a gap which many people have felt existed, but which has not apparently been defined in print before. Some officers have of course become aware of what is involved as a result of their experiences over a period of time, but that is a very different thing from being taught about it in advance and thereby being in a position to adapt the idea to the circumstances of a particular situation from the start. Of these two recommendations the second is certainly more important to the United

States Army because a unified establishment for counter-insurgency specialists has already been set up.

Under certain circumstances a third recommendation might become as important as the two already mentioned. If the analysis given and the predictions made in Chapter 1 are correct, the second half of the 1970's is going to see a further swing towards the lower end of the operational spectrum with large scale insurgency giving way to civil disorder accompanied by sabotage and terrorism, especially in urban areas. If this happens the operational emphasis will swing away from the process of destroying relatively large groups of armed insurgents towards the business of divorcing extremist elements from the population which they are trying to subvert. This means that persuasion will become more important in comparison with armed offensive action, although both will continue to be required, and both will be equally dependent on good information. In terms of preparation, the effect of this is to enhance the priority which should now be given to teaching officers the methods of carrying out large scale persuasion, and to providing the Psychological Operations specialists and units which will be required.

But one thing stands out regardless of priorities, which is that the once mysterious processes used by the organizers of subversion to turn a section of a people against its government, have long ago been exposed for all to see. Many of those who have practised subversion have written about their ideas and experiences, and other writers have not been slow in interpreting them. Anyone can discover all that they could possibly want to know, by reading books which are sold on the open market and held in libraries throughout the world. Furthermore the same situation exists with regard to the measures which have been found successful for countering subversion and insurgency, with the exception of the tactical handling of information by operational commanders. Unfortunately the fact that so much is known by those who have studied the problem does not mean that the problem has been studied by all those who should have done so. In practice a considerable number of officers, including some senior ones, still consider that it is unnecessary to make any great effort to understand what is involved, and the cry that a fit soldier with a rifle can do all that is required is frequently heard to this day. It could be argued that more would be achieved by eradicating this attitude than by implementing all the recommendations of this book and regardless of whether or not this is literally true, it would obviously be of great benefit to do so. Unfortunately the attitude is deeply ingrained in regular officers on both sides of the Atlantic. The reasons for this are highly complex and a separate study would be required to do justice to them. It might however be worth concluding this one by looking at the main factors involved. Although such a cursory examination can not be expected to lead to any fundamental conclusions, it might at least serve to provide a stepping stone for a further study.

At the root of the problem lies the fact that the qualities required for fighting conventional war are different from those required for dealing with subversion or insurgency; or for taking part in peace-keeping operations for that matter. Traditionally a soldier is trained and conditioned to be strong, courageous, direct and aggressive, but when men endowed with these qualities become involved in fighting subversion they often find that their good points are exploited by the enemy. For example firm reaction in the face of provocation may be twisted by clever propaganda in such a way that soldiers find the civilian population regarding their strength as brutality, and their direct and honest efforts at helping to restore order as the ridiculous blunderings of a herd of elephants. Gradually the more intelligent officers find themselves developing a new set of characteristics such as deviousness, patience, and a determination to outwit their opponents by all means compatible with the achievement of the aim. Those who are not capable of developing these characteristics are inclined to retreat into their military shells and try not to notice what is going on. They adopt the 'fit soldier with a rifle' theory, and long for the day when they can get back to 'proper soldiering' by which they mean preparing for the next – or last – war, as opposed to fighting in the current one. There are also some sound material reasons for not becoming too well qualified in fighting insurgents, because expertise in this field can result in an officer being channelled away from the main stream of military preferment, a factor which is more apparent in the United States Army than in the British Army.

Although it is comparatively easy to understand why officers are reluctant to become immersed in the problems of handling subversion, it is less easy to know what should be done about it. It is of course impractical to suggest that officers should be told to think in a totally different way about military matters, because the cumulative effect of centuries of conditioning can not be discarded so easily. In any case the army is not solely maintained to fight insurgents or to take part in peace-keeping operations. In practical terms the situation may improve to some extent as the better educated junior officers replace those who are retiring, and any improvements made in the education and training given to officers will be reflected in this way. Considerable progress in adjusting the attitudes of those actually employed in countering subversion and insurgency could be achieved by careful briefing of those commanding units. At the moment many of these people deliberately try to present the situation to their subordinates in terms of conventional war. They make rousing speeches about knocking the enemy for six, and they indulge in frequent redeployments and other activities designed to create the illusion of battle. But quite apart from the tactical disadvantages which accrue, e.g. lack of continuity, they actually manage to aggravate the strains on their subordinates because they are in effect encouraging the development of the characteristics which are unsuited to this particular type of operation, whilst retarding the growth of those which might be useful. In other words they are leading their men away from the real battlefield onto a fictitious one of their own imagining. Instead, commanders would be better employed in explaining the fundamental realities of the situation to their subordinates and in encouraging them to submerge themselves in the atmosphere of the country. Only by so doing will they be able to see things from the point of view of the population whose allegiance they are trying to regain and retain.

The question of men's attitudes is an interesting one but although in a sense relevant to the purpose of this study, it is so hedged around with imponderables that no useful purpose would be served by further speculation in this context. Perhaps some qualified person will take the matter up later on, and research it in a scientific way. Meanwhile it is permissible to hope that the contents of this book will in some way help the army to prepare itself for any storms which may lie ahead in the second half of the 1970's.

Bibliography

The following list contains all books referred to in the footnotes together with a number of others which were found to be particularly useful in the preparation of this study. Unpublished works, and articles contained in newspapers and periodicals are not included.

ALI, TARIQ, *New Revolutionaries, Left Opposition*, Peter Owen, 1969.
BARKER, DUDLEY, *Grivas, Portrait of a Terrorist*, Cresset Press, 1959.
BARNETT, D. and KARARI NJAMA, *Mau Mau From Within*, Macgibbon and Kee, 1966.
BLACKSTONE, GALES, HADLEY and LEWIS, *Students in Conflict*, Wiedenfeld and Nicolson, 1970.
BUCHAN, ALISTAIR, *Europe's Futures, Europe's Choices*, Chatto and Windus, 1969.
BYFORD JONES, W., *Grivas and the Story of Eoka*, Robert Hayle, 1959.
CALDER, *Unless Peace Comes*, Alan Lane, The Penguin Press, 1968.
CAMPBELL, ARTHUR, *Guerillas*, Arthur Barker, 1967.
CHALIAND, GERARD, *Armed Struggle in Africa*, Monthly Review, 1969.
CLARKE, ROBIN, *We All Fall Down*, Alan Lane, The Penguin Press, 1968.
CLUTTERBUCK, RICHARD, *The Long Long War*, Cassell, 1967.
COHN BENDIT, D., *Obsolete Communism, The Left Wing Alternative*, André Deutsch, 1968.
CRITCHLEY, T. A., *The Conquest of Violence*, Constable, 1970.
DEBRAY, REGIS, *Revolution in the Revolution?*, Pelican Books, 1968.
DEBRAY, REGIS, *Strategy for Revolution*, Jonathan Cape, 1970.
DIXON and HEILBRONN, *Communist Guerilla Warfare*, Allan and Unwin, 1954.
FISHMAN, W. J., *The Insurrectionists*, Methuen, 1970.
FITZGIBBON, CONSTANTINE, *Out of the Lion's Paw*, Macdonald, 1969.
FOLEY, CHARLES, *Island in Revolt*, Longmans, 1962.
GREIG, IAN, *Assault on the West*, Foreign Affairs Publication House, 1969.
GRIVAS, GEORGE, *Guerilla Warfare, Eoka's Struggle*, Longmans, 1964.
GRIVAS, GEORGE, *The Memoirs of General Grivas*, edited by Charles Foley, Longmans, 1964.
GUEVARA, ERNESTO, *Guerilla Warfare: A Method*, Normount Armament Company and Co., 1962.
GUEVARA, ERNESTO, *On Guerilla Warfare*, Cassell, 1962.
GUEVARA, ERNESTO, *Bolivian Diary*, Jonathan Cape, 1968.
GUEVARA, ERNESTO, *Reminiscences of the Cuban Revolutionary War*, Pelican Books, 1969.
HARBOTTLE, MICHAEL, *The Impartial Soldier*, Oxford University Press, 1970.
HASTINGS, MAX, *Ulster 1969*, Victor Gollancz, 1970.
HEILBRUNN, OTTO, *Warfare in the Enemy's Rear*, Allen and Unwin, 1963.
HENDERSON and GOODHEART, *The Hunt for Kimathi*, Hamish Hamilton, 1958.
HOCH and SCHOENBACH, *The Natives Are Restless*, Sheed and Ward, 1969.
HUCK, ARTHUR, *The Security of China*, Chatto and Windus, 1970.
JAMES, ALAN, *The Politics of Peace Keeping*, Chatto and Windus, 1969.
JAMES, ROBERT RHODES, *The Czechoslovak Crisis*, Wiedenfeld and Nicolson, 1969.
KITSON, FRANK, *Gangs and Countergangs*, Barrie and Rockliff, 1960.
LAWRENCE, T. E., *Seven Pillars of Wisdom*, Jonathan Cape, 1935.
LEAKEY, L. S. B., *Mau Mau and the Kikuyu*, Methuen, 1952.
LEAKEY, L. S. B., *Defeating Mau Mau*, Methuen, 1954.
LEGAULT, ALBERT, *Peace-Keeping Operations: Bibliography*, IPKO Publications, 1967.
LONNROTH, ERIK, *Lawrence of Arabia*, Valentine Mitchell, 1956.

LUTTWAK, EDWARD, *Coup D'Etat*, Allen Lane, The Penguin Press, 1968.
MAJDALANEY, F., *State of Emergency*, Longmans, 1962.
MAO TSE TUNG, *On Guerilla Warfare* Translated by S. Griffiths, Cassell, 1962.
MCCARTHY, RICHARD, *The Ultimate Folly*, Victor Gollancz, 1970.
MCCUEN, JOHN, *The Art of Counter-Revolutionary War*, Faber and Faber, 1966.
MIKSCHE, F. O., *Secret Forces*, Faber and Faber, 1950.
MITCHELL, COLIN, *Having Been a Soldier*, Hamish Hamilton, 1969.
MUUS, FLEMMING, *The Spark and the Flame*, Museum Press, 1956.
O'BALLANCE, EDGAR, *The Indo China War*, Faber and Faber, 1964.
O'BALLANCE, EDGAR, *The Greek Civil War*, Faber and Faber, 1966.
O'BALLANCE, EDGAR, *The Algerian Insurrection*, Faber and Faber, 1967.
OPPENHEIMER, MARTIN, *Urban Guerillas*, Penguin, 1970.
PAGE, LEITCH and KNIGHTLY, *Philby*, André Deutsch, 1968.
PAGET, JULIAN, *Counter-Insurgency Campaigning*, Faber and Faber, 1967.
PAGET, JULIAN, *Last Post : Aden 1964–67*, Faber and Faber, 1969.
PHILBY, KIM, *My Silent War*, Macgibbon and Kee, 1968.
RIKHYE, I. J., *United Nations Peace-Keeping Operations: Higher Conduct*, IPKO Publications, 1967.
ROBERTS, ADAM, *The Strategy of Civilian Defence*, Faber and Faber, 1967.
ROY, JULES, *The Battle of Dien Bien Phu*, Faber and Faber, 1965.
SINCLAIR, ANDREW, *Guevara*, Fontana, 1970.
SKEM, L. M. K., *Military Strategy at UN Headquarters For Peace Keeping Operations: A Proposal*, IPKO Publications, 1967.
STENGUIST, NILS, *The Swedish UN Stand By Force and Experience*, IPKO Publications, 1967.
STEPHENS, R., *Cyprus, A Place of Arms*, Pall Mall Press, 1966.
STRONG, KENNETH, *Intelligence at the Top*, Cassell, 1968.
SUN TZU, *The Art of War*, Translated by S. Griffiths, Oxford University Press, 1963.
TABOR, RICHARD, *The War of the Flea*, Paladin, 1970.
TANHAM, GEORGE, *Communist Revolutionary Warfare*, Frederick Praeger, 1961.
THOMPSON, ROBERT, *Defeating Communist Insurgency*, Chatto and Windus, 1967.
THOMPSON, ROBERT, *No Exit from Vietnam*, Chatto and Windus, 1969.
TRINQUIER, ROGER, *Modern Warfare*, Pall Mall Press, 1964.
VALERIANO AND BOHANNAN, *Counter Guerilla Operations: The Philippine Experience*, Pall Mall Press, 1962.
VON HORN, CARL, *Soldiering for Peace*, Cassell, 1966.
VO NGUYEN GIAP, *Big Victory, Great Task*, Pall Mall Press, 1968.
VO NGUYEN GIAP, *Peoples War, Peoples Army*, Bantam, 1968.
WILSON, A. J., *Some Principles for Peace-Keeping Operations: A Guide for Senior Officers*, IPKO Publications, 1967.
WOODCOCK, GEORGE, *Civil Disobedience*, CBO Publications, 1966.
WOOLMAN, DAVID, *Rebels in the Rif*, Oxford University Press, 1969.

Index

205

INDEX

Dominican Republic, 146
Donnison, David, 83 n. 2
Dubcek, Alexander, 22

Egypt, 19, 22, 146
Encylopaedia Britannica, 14
ENOSIS, 33
EOKA, 17, 30, 127, 132
Erskine, General, 135
Europe's Futures, Europe's Choices (ed. A. Buchan), 23 n.

Fertig, Colonel, 73
Finland, 144
Foco theory, 33–4, 39, 42
Foley, Charles, 33 n. 2, 100 n. 5
Fraternizing dangers, 91

Galtung, John, 16
Gandhi, Mahatma, 82
Gangs and Counter Gangs (Kitson), 31 n. 4, 6, 123 n.
Gas, use of non-lethal, 139–40
Greek civil war, 40, 57–8
Greek Civil War, The (O'Ballance), 40 n. 1–3, 58 n. 1–2
Greek Cypriots, 30, 147–8, 150
Greig, Ian, 16 n. 3, 18 n. 1, 22, 38 n. 1–2
Grivas, General George, 2, 5, 33, 40 n. 9, 100 n. 5, 132
Grivas, The Memoirs of General (Grivas and Foley), 33 n. 2, 100 n. 5
'Guerilla and Revolutionary War', 170 n.
Guerillas (A. Campbell), 31 n. 1, 32 n. 2, 51 n., 135 n.
Guerilla war, 2, 5, 21–3, 32, 39–43, 46, 94
defined, 14–15
pre-Christian origins, 15
Guerilla Warfare: A Method (Guevara), 33 n. 5
Guerilla Warfare (Grivas), 5 n. 2–3, 40 n. 9, 132 n.
'Guerilla Warfare, Changing Pattern of' (de la Billiere), 21 n. 1
Guerilla Warfare (Mao Tse Tung and Che Guevara), 15 n. 2, 16 n. 1, 21 n. 4
Guevara, Che 15 n. 2, 16 n. 1, 29, 33–4, 42
Guevara (Sinclair), 34 n. 3
Guinea, Portuguese, 35–7

Hammarskjold, Dag, 26
Harbottle, M., 150 n. 2, 152 n. 1–2, 155 n., 157 n. 1–2
Harding, Field-Marshal, 57
Having Been a Soldier (Mitchell), 90 n.

Heilbrunn, Otto, 45 n., 46
Helicopters, dubious value in countering insurgency, 137–9
Hoch, P., 83 n. 1, 87, 88 n. 1
Holland, 145
Hong Kong, 173
Horn, General von, 152–3, 157, 160

Image intensifiers, 141
Impartial Soldier, The (Harbottle), 150 n. 2, 152 n. 1–2, 155 n., 157 n. 1–2
India-Pakistan conflict, 18, 145
Indo-China, 30
Indo China War, The (O'Ballance), 30 n. 2–3
Indonesia, 1, 145, 185
Infantry, School of, 168, 171–3
Information, 95–101, 126–31
build-up, 97–101
handling, need for deeper training in, 173, 199
Institute for Strategic Studies, 23
Insurgency
causes behind recent, 29–38
definition, 3
Insurrection, hypothetical; background, factual, 102–26
Intelligence, 7–8, 39, 59, 71–8, 81, 91–2, 95–101, 129, 131, 188–92, 194–6
interrogation, data links and, 142–3
operational, 72–3, 76
organizations unpopular during peace, 71–2
political, 72–3, 75
source, police as, 76

James, Alan, 144 n. 1–2, 145 n. 1–5, 146 n. 1–5, 149 n. 1–4
Jerusalem, 38
Joint Services Staff College, 173 n. 1, 176
Joint Welfare Establishment, 188

Kashmir, 25
Katanga, 149
Kenya, 31, 33, 42, 100, 123, 135–6
Kikuyu, 31
Kitson, F. E., 31 n. 4, 6, 123 n.
Korea, 1, 46, 146
Kuwait, 1

Last Post: Aden 1964–67, (Paget), 17 n. 2, 38 n. 3, 50 n., 57 n. 1, 76 n. 5, 136 n.
Lawrence, T. E., 21
League of Nations, 144
Leakey, L. S. B., 31 n. 2–3
Legault, Albert, 167 n. 1
Lenin, Vladimir, 32
Liddell Hart, B. H., 16, 91

206